MARS MISSION I

SURVIVING THE KESSLER EFFECT

CHRISTOPHER LEE JONES

MARS MISSION I

This is a work of fiction. Names, characters, places, historical events and incidents are the product of the authors imagination. Every effort has been made to adhere to scientifically verifiable data and facts. It is still at least at the moment, a work of fiction.

Published by

On the Hook Publishing, LLC

U.S.A

ISBN # 978-1-7342246-0-3 (Paperback)

ISBN # 978-1-7342246-1-0 (Hardcover)

Library of Congress Control Number: 2020903798

Cover design: German Creative

Printed in the United States of America

www.christopherleejones.com

Dedication

This book is dedicated to Elon Musk, who has inspired me through his incredible work ethic, his dedication to first principles physics, the simplest and most sane way to look at the universe we all inhabit, and the incredible number of success' he has accomplished through Spacex and Tesla. I am very appreciative of his dedication to help mankind return our behaviors and actions back into a sustainable method of living and his efforts to help us turn humans into a multi-planetary species and explode our imaginations. Back in 2017 at the Kennedy Space Center, I was able to witness one of the first Falcon 9 missions to return its booster safely back to Kennedy Space Center. I was unprepared for what I witnessed and realized immediately that I had just seen an amazing accomplishment of engineering and the future of space. At that moment the hibernating space nut in me was re-awoken. I have since witnessed the first Falcon Heavy launch, the one with Elon's Red Roadster inside, when the two boosters returned to land successfully. I have travelled to many different NASA destinations including the United States Space and Rocket Center in Huntsville, Alabama, the Red Stone Rocket Center and Marshall Space Flight Center on different occasions which was also the previous home of Werner Von Braun. Who was arguably one of the first inventors of rockets in history. Without him the Apollo era and the humongous Saturn V, still the largest and most powerful rocket in history, would have never been realized. And NASA's missions in the late 1960's and early 1970's and the Apollo landings on the moon would have never happened.

I have to agree with Elon that humans must have imagination to live. They must have a reason to get up in the morning to survive and a focus for which to strive. He has re-awoken that in me. I believe I will live to see humans walking on the surface of Mars, due to him. I just hope we do not destroy our only way off of this planet through low Earth orbital debris and thereby awakening the Kessler Effect.

I am very grateful to Donald Kessler for his words of support and his thoughtful endorsement as well as his letter to the reader.

I am grateful to my Mother, Marvel Jones a typo and spell check champion and my Father, Ronald Jones without whose support this book would have never come to fruition.

I am also grateful to Gabriela Kilianski who has spent many days and evenings going over the details of this book with me and its organization and listening with interest as I repetitively discussed aspects of this story.

I am eternally grateful to Sarah Flores-Write Down the Line, LLC who ended up spending a significantly extended period of time trying to make sense of my original gibberish, run-on thoughts, incomplete sentences, and generally poor writing style to turn this into a work that without her would have never been published.

A Letter from Donald Kessler

Mars Mission I is a fictional story of humans' first landing and walk on Mars, an event that would be celebrated even more than our first landing on the Moon....but only if the world was aware of this success and didn't face other issues arising from what is sometimes called "The Kessler Effect" first predicted in 1978.

The Kessler Effect (also referred to as the Kessler Syndrome) is a phenomenon where objects in Earth orbit eventually collide with one another, where each collision produces a cascade of orbital debris, causing more collisions and debris. The rate of these collisions increases exponentially with the number of collisions and the amount of mass in Earth orbit, with both contributors having a history of continually increasing. The increase continues despite the adoption of international guidelines enacted to prevent it.

Earth orbital space is used by the military, civilian, scientific and commercial communities, with each having different objectives. It has been known for some time that if a major war were to occur, military satellites would be the first subject of aggressive action. In 1985, the US military planned an anti-satellite (ASAT) test and against NASA's recommendation, launched a rocket in a trajectory to collide with a satellite at the orbital speed of 17,000 miles per hour. The military had not believed NASA's predictions. In 2017, the Chinese conducted an ASAT test without consulting their own scientific community, which would have advised against it. And what did each nation learn from these tests? They learned that they should find a different method of neutralizing enemy military satellites because of the devastating consequences to the space environment.

This book accurately reflects our current understanding of orbital debris issues and introduces another debris issue that has not been fully developed by the scientific community. These facts are combined into a cliff-hanging detective story of what a country might do to gain an advantage over a perceived enemy and becomes a story of survival that humanity may face if we are unable to do what we already know needs to be done.

Donald Kessler

Retired NASA Senior Scientist for Orbital Debris Research

MARS MISSION I

SURVIVING THE KESSLER EFFECT

1. That's Not a Piling!

Bob and Brian grew up in North Miami. Born in the 1960s, they saw a lot of changes to South Florida over the years. They had been fishing buddies since the 1980s, and now that they were both retired from the fire department, they had even more time for fishing.

Brian retired first after picking up a patient in an awkward position, causing him back strain that continued to fester and then turned into a neck issue over time. The neck pain slowly started causing sleep disruption, and then brain fog caused him memory and mood issues. It began to negatively affect his quality of life, so it was time to call it a day before depression set in.

Being forced into retirement can be devastating, but Brian knew it was time, so he left after eighteen years—before things got worse. The only thing they did together besides fish was watch the rockets launch out of Cape Canaveral. They had been to Kennedy Space Center to watch the launch of all ten of the humongous starships, including the beautiful *Starship X* that launched six months before, with its crew of twenty-one courageous astronauts. This was still exciting for them because *Starship X* was now landing on Mars.

Bob had set his home entertainment system to record the landing so they could watch it together over a few beers. Bob retired after twenty-four years, and now the two of them were free to fish as much as they wanted, and they enjoyed it immensely. They were known by the local fisherman as the B&B twins, and at every opportunity they were out fishing, even if the wind, weather, and seas were not in their favor.

It was early August, and the summer trade winds had let up enough to give them a gorgeous day 20 miles off Fort Lauderdale in the Gulf Stream. The seas were almost flat with a slight breeze out of the southeast. The dolphins were missing more than hitting, but that didn't really matter. The sea was a beautiful sterling blue with tiny wind-driven waves walking their way from who knows where.

Things aboard the sturdy little craft were unusually quiet when Brian thought he saw something unusual. It looked like a big channel marker that had no reason being so far out to sea. But even more than that, something was wrong with it. It was bobbing up and down and it looked too big. A broken piling wouldn't bounce up and down like a child in a bounce house or on a trampoline. Brian stood up on the bow to get a better look at the thing bobbing in the water about a mile away. Suddenly, a huge spray of steam that sounded like

11

a jet engine taking off, accompanied by billowing smoke, sea spray, and total chaos erupted off the port side bow. It was so loud Brian lost his balance backward as if a bomb had gone off right in front of him. A geyser of water and fire rocketed upward with a deafening, thunderous roar.

They both immediately dropped their poles. At first, Bob thought it was a plane crash as he saw Brian fall back and smack his head on the helm. Blood flowed from the back of Brian's head as he writhed around, holding it with his hands and screaming, but he could not be heard over the incredible noise as Bob watched a flaming rocket disappear into the sun. Bob watched it as long as he could using his sunglasses and hat to shield his eyes from the intense light, but he needed his hands to cover his ears and he couldn't do both at the same time. The noise was so loud, it had to be a bomb. But it was moving. It was so ear shattering, he couldn't think at all or even imagine what could possibly produce such a horrendous volume of sound. Then the blast wave hit him, and he fell to the floor of the small boat.

The "anomaly," as it was first immediately identified, came up almost simultaneously on both the Fort Lauderdale and Miami Airport radars. A few seconds later, the "anomaly" was being tracked on the Homestead Air Force

Base radar. Seconds after that, it was spotted on the joint Naval Air Force radar out of Key West. Alarms were immediately triggered, and two F-16 fighters were dispatched out of Homestead Airforce Base with four more dispatched out of Key West. Calls immediately went out to NASA to see if they had an unscheduled launch. The strange thing was, whatever this anomaly was, it was not "inbound," so it didn't appear to be a threat—at least not to the U.S. mainland. The response was rapid, and tracking confirmed that whatever it was, it was going straight up.

2. *Starship X*

It was 5:55 a.m., February 14th, and the world was getting a gigantic valentine. *Starship X* was vertical on Launch Pad 39A, all 387 feet gleaming in the xenon lights as the horizon was turning pink with the first rays of the morning sun. The stainless-steel-clad, nine-million-pound rocket was breathtakingly beautiful and could be seen from as far as 10 miles away. The twenty-one astronauts comfortably strapped in their custom high-back chairs were quietly listening to the countdown's progress. The cockpit, lit only from the light of the touchscreen monitors, provided a dark, cool cocoon for the astronauts to focus on the work at hand. The only noise was the sound of the fuel topping up the groaning tanks as the supercooled propellants caused the metal tanks to contract.

Mission Specialist Andrea Tripp from Houston, Texas, thought the starship sounded like a giant waking from a deep sleep, as Commander Peter Carrigan and Chief Pilot Mary Pfeiffer were finishing up their pre-launch checklists. Communications Specialist Steven Henderson was confirming the checklists were complete, as the countdown progressed under twenty seconds.

Slowly, the monitor lighting grew in intensity as if the stage were set for the show to begin. The orchestra was all seated. The maestro picked up the baton. The audience went silent as the astronauts heard in their headsets, ". . . nine, eight, seven . . . main engines, start!" The groaning disappeared into a roaring thunderclap that could not be stopped. ". . . six, five, four, three, two, one . . . liftoff!"

Nine million pounds of thrust slammed the 34-story rocket upwards, using 40 tons of supercooled methylox fuel per second. *Starship X* was shaking and banging worse than anyone on board had imagined. As the acceleration increased, the increasing g-forces took control. All twenty-one astronauts were forced deep into their seats and headrests. The touch screen monitors were now virtually unreadable due to the increasing vibrations. A voice in their headsets sounded. "*Starship X*, go for roll."

Commander Carrigan responded with a barely audible yell. "Roger, roll."

As the roll program gimbaled the giant engines and thrust to the correct trajectory for low Earth orbit, the gleaming ship changed course and heading. *Starship X* was no longer talking to Cape Canaveral; they were now talking with

CAPCOM in Houston—if they could hear them over the roar of the engines.

"*Starship X*, go with throttle up," was the command from CAPCOM.

"Roger, throttle up," said Commander Carrigan.

Andrea looked at her hands on the armrests. They had trained for this, but experiencing it was out of this world. She tried to raise her left hand. It took all of the energy she had to pick it up six inches, battling the ever-increasing g-forces.

This is crazy, she thought.

They were now like twenty-one leaves of a tree being violently shaken in a vertical hurricane of uncontrollable proportions. *Was this "normal*?" All twenty-one astronauts appeared to be dolls in an angry child's toy. They were thrown left, then right, then forward. Thank God for the five-point harnesses and the custom-made, high-impact seating. They were vulnerable to this gigantic metal tube having its way with their bodies. Did the engineers get everything right? How could this enormous structure that was to be their taxi and home for the next two years possibly hold together under this stress? Andrea was sure *Starship X* was going to break apart, then it happened. Something really big.

Nothing.

Andrea's eyes told her that she wasn't in heaven but she was close. *Either nothing is noisy, or we all just died,* she thought.

They had achieved weightlessness. The blue Florida sky was now the impossible infinite black of space.

3. Starships

Starships I and *II* launched without a hitch, giving everyone involved and all space fans high hopes. Two years went by while *Starship III* was readied. During countdown, a glitch kept reappearing, showing that a sensor was faulty. After investigation, it was launched without further incident. *Starship IV*, like *I* and *II*, went off without any issues, as did *Starships V*, *VI*, and *VII*.

Starship VIII prelaunch progress continued smoothly until the final twenty-two seconds to launch, when a pressure switch on one of the huge fuel tanks had to be replaced. To fix the problem, they spent two days removing all 6 million pounds of fuel before refueling. *Starship IX* also launched as expected with no anomalies. These Starships left launch pad 39A out of Cape Canaveral, Florida, approximately three months before *Starship X*'s launch with its precious human cargo and support supplies. The eyes of the world had watched as *Starship X* launched flawlessly and climbed into space on its six-month journey to Mars.

4. Mars Mission I

As soon as they entered low Earth orbit, they had to rendezvous with a tanker to refuel. This required virtually no work on their part as it was completely automated. A tanker was waiting for them intentionally in the shadow of Earth to keep the fuel cool. Cold fuel means more fuel at a higher density. Commander Carrigan and Chief Pilot Mary Pfeiffer monitored the operation. Before long, *Starship X* had a half million pounds of methane and almost two million pounds of liquid oxygen back in her tanks.

CAPCOM issued the "go" order for Mars Injection Orbit, and the twenty-one astronauts felt gravity return as *Starship X* throttled up for the thirty-five-million mile, six-month-long journey to Mars. As soon as *Starship X* was en route on the correct trajectory, the crew began referring to their massive ship as "*Staten*," in place of *Starship X*. Andrea had commented to someone, "It's just easier to say." so it stuck. The sleeping rotation, food preparation and exercise regimen were initiated as planned. They were now under way. Next stop Mars.

The crew consisted of men and women from seven different countries. This was intentionally an international effort, and every astronaut candidate had been thoroughly vetted and drilled to the point of exhaustion. Personality and character testing had been conducted by psychologists, psychiatrists, and human behaviorists, with the intention of finding the slightest expression of simple frustration, personality flaw or prejudice, and if expressed, they were quickly dismissed. Even the most patient people had a hard time with the redundancy and monotony of the testing regimen.

By the end of testing, the final candidates were picked because they simply didn't show frustration or boredom. In the end, the successful astronauts understood this as if it were a deep, personal secret, and of course, it was never discussed because each one of them wanted the best character and personality traits to be on the mission as well. There were no inside secrets or shortcuts. After all, this was going to be the longest mission in history. A key factor to many of the winning personalities of the astronaut candidates (also known as "Ascans") was that they always knew they would be the first to land on Mars—or at least among the first and anything would be worth that.

"Hey, who cares if I'm on the first or the fifth starship to go? How many people can say they went to Mars?" said Pedro Lopez to the *Miami Herald* during an interview.

Pedro was the flight engineer tasked with assisting Commander Carrigan and Mary as well as relating data based on position verification and instrumentation verification and accuracy with CAPCOM (originally meaning Capsule Communication), out of Houston Ground Control.

Everyone on board was thoroughly cross-trained and routinely sat in different seats, performing many different duties. That is, all except Commander Carrigan, whose job description was basically God Junior on *Staten*—a position few people could ever imagine, let alone handle.

Once the final engine burn achieved its flight heading and speed objective, zero gravity returned, and the astronaut shift rotation was put in place. The three shifts, referred to as A, B, C, or Alpha, Bravo, and Charley and they broke the twenty one crew members into two shifts of seven astronauts and one of six. Within a couple of days, the shifts were all functioning as expected, and each person was doing his or her part. In less than a week, the ship and crew were functioning so smoothly and cohesively that Commander Carrigan had nothing to do other than verify that the ship's log entries were correct, thorough, and timely and his commander's log entries were the same.

Commander Carrigan often found himself working out in the gym on the circular treadmill, specifically designed for zero G's, or on the weight training equipment. It was one of the best places for him to focus on the job at hand and stay sharp for what may come to pass. Every time he checked on technical operations or the more monotonous duties, they were being taken care of correctly. He even checked in with the flight surgeon to make sure everyone was still checking out correctly with their weekly health check-up. Each time he evaluated the big ship's operations, course, heading, fuel, oxidizer, radiation exposure, engines, tanks, or even the galley for cleanliness or general sanitation, everything was operating as expected. It was as if they had been there before, and well, they had. That's what the months and years of drilling and training had been all about. What was not expected was that after the first three months of travel, everyone was genuinely enjoying the journey and having a good time.

As to be expected, the three shifts were rotated bi-weekly so everyone on board had an opportunity to work with everyone else, and no one fulfilled the same duties for weeks on end. This rotation was done so everyone stayed familiar with different duties, reducing the likelihood of boredom and the tendency of tribalism and territory or kingdom creation.

This human phenomenon of territory claiming can easily be witnessed in many businesses and government offices today. When people clutter their desks and workspaces with a multitude of personal pictures and paraphernalia, they are effectively saying, "This is my territory. Do not touch." No one person "owned" any of the public spaces aboard *Staten*, and everyone knew it. There were plenty of public spaces. The helm and communications, had an out of this world observatory at the nose or bow, below the helm was the astronaut safety seating for launches and landings. The "day room." below the astronaut safety seating lead to the galley and below that was the gym with its thirty-foot circular treadmill for use in zero gravity, hydraulic weight machines, core machines, and bungee cords. The showers were below the work out room, which you could enter through a door straight from the gym. To conserve water, and since much of the trip would be zero gravity, a shower consisted of wrapping the showerhead or spicket with a towel, wetting it thoroughly, and wiping themselves down with the wet towel. They would then follow it with a "soap towel," which was the size of a washcloth and had soap infused into the fabric. This was finally followed with the wet towel again to get the soap off.

Towels were a huge asset and were to be used sparingly. To save space they were vacuum packed as tight as bricks until opened and removed from their packing material. They had six of their own towels issued in their bunk rooms. These six towels were to last until arrival on Mars. Then they were issued three new towels. The objective was to use the fewest towels possible. Storage space and weight were a premium, and the longer they could make things last, the more room they had for fuel and food. Most used one towel at a time until they couldn't bear to use it any longer.

The medical room was right below the showers and consisted of everything found in an emergency medical clinic: an x-ray machine, cardiac equipment, tools for minor and some major surgeries, as well as trauma supplies. The computer room was below the clinic and provided a quiet space for work and personal email. The surround sound entertainment room, which was virtually soundproof when the double doors were closed, had over ten thousand of the most recent and golden oldie movies. Below the

24

entertainment room was storage space, and below that began the personal bunk rooms.

Each crew member had a personal space, or bunk room, as it was identified in the Policies and Procedures manual of Mars Mission I (MMI). Each bunk space was 8 feet long by 6 feet wide, with a 7-foot ceiling. In zero gravity, this was fairly spacious and luxurious. Each bunk room had a privacy door and places to put clothing and personals. It was also outlined in the manual that the ship would be returned in "as new condition" as possible upon arrival back in Cape Canaveral. It was entirely possible the ship would be readied immediately for another mission, and any "unusual cleaning services" (whatever that meant) would be charged to the astronaut's paycheck. So, the manual suggested using discretion in adorning personal spaces. It also suggested that they are expected to return to Earth.

Mars was rapidly growing in the windows, and it was to no one's surprise when Pedro announced what everyone else knew: that *Staten's* trajectory was "dead on" and projected perfectly for an inbound landing on Mars's massive canyon, the Valles Marineris. It still brought a round of applause and smiles.

As expected, the shift rotation was immediately canceled for the landing. All astronauts took their personal seats, which were conformed specifically to their bodies while wearing their Mars suits. They tightened the five-point harnesses that held them firmly in place. They didn't have to go "on air" unless a specific warning sounded, they were instructed to, or, of course, the situation required.

It was dead quiet on board as *Staten* slowly rotated and adjusted its heading. The attitude thrusters kicked in and a slight rushing of air was heard followed by some vibrations which began to grow. The sound of rushing air continued to increase as did the vibrations. As *Staten* encountered the Mars atmosphere the vibrations turned to a bronco busting ride. The retro boosters fired, and gravity slowly increased. On approach to Mars, depending on the angle of attack, gravity could approach as much as 5 g's, or five times a human's weight on Earth. An astronaut of 150 pounds (68 kg) could experience as much as 750 pounds (240 Kg) of gravity.

The landing on Mars is designed to be completely automated and choreographed by the navigational computer, with inputs from 274 sensors, the onboard radar, as well as active, real-time cameras and lidar sensors. All thrusters were operational and continued to control the gigantic, 180-foot-tall by 33-foot-wide spaceship. Chief Pilot Mary Pfeiffer sat at the controls monitoring every aspect of the approach. As expected, the noise and bucking decreased rapidly.

The crew felt a slight bump as the engines cut out. Anxious eyes looked around from one astronaut to another. It was silent.

The distance from Earth to Mars is 34 million miles at its closest and 140 million miles at its farthest, depending on the positions of their orbits. Once a year, Mars and Earth are on different sides of the sun, making communication difficult if not impossible. Fortunately, different satellites help to "deflect" the transmissions around the curvature of the sun. It takes between three and twenty-two minutes for radio transmissions to travel that distance one way. Houston would be of little help. It is a six-month trip with a required one-year stay and another six months back home. As one flight attendant watching the launch remarked, "That's a long layover!"

Commander Carrigan shouted over the PA. "Valles Marineris, everybody!"

The tenth Starship landed safely, and this one had humans on board. For the first time in history, humans were not only on Mars, but on another planet. Commander

Carrigan and everyone else felt the effects of the six-month journey in zero gravity. The approximate 36% g, or 36% of Mother Earth's gravity, felt like two or three times that of Earth's immediately upon landing.

"It's in our training to expect this situation," Mary said, "but reality still sometimes sucks."

"You're not kidding," Commander Carrigan replied. "I feel like I was either out drinking all night or we landed on Jupiter."

A huge collective sigh of relief was heard as seatbelts were unbuckled and the highly trained astronauts started adjusting to the new gravity. Some had a little nausea as a result of the sudden onset of gravity, and one or two were afraid to stand for a few minutes. Some of them were dizzy as well. The crew had been trained to expect this, and they were given time to acclimate. A little at a time, they were returning to some kind of homeostasis and a growing feeling of "normalcy". As if the first humans on Mars were going to feel "normal" for some time. Still, time was given for them to slowly adjust.

The chief briefing en route described the status of the other nine starships. Everything needed for the two-year round trip that they weren't already carrying on *Staten* was sent in advance on the other nine starships. CAPCOM informed Commander Carrigan, who then informed the crew, that *Starship IV* had fallen over after landing, according to NASA. Since every starship had duplications and redundancies, the mission was still a go. One of their first assignments would be to assess the situation with *Starship IV*. The other eight starships were performing nominally.

Starships *I* and *II* had been sent two years in advance. They were actively producing water from the atmosphere, as well as methane for fuel, utilizing hydrogen fuel cells and solar cells for power. A special, remotely-operated piece of equipment had been autonomously off-loaded as well. For the past two years, it had been making methane utilizing the "Sabatier method" and safely storing it in one of the onboard tanks, which was about half full by the time *Starship X* launched from the now-world-famous Launch Pad 39A. This guaranteed that enough fuel would be immediately available for their return trip since none of the starships carried enough fuel on their own for the 35-million-mile return trip back to Earth.

Captain Carrigan assembled the team for a quick situational briefing. He knew this wasn't really required for the crew, but he understood too well that if he didn't assemble the troops for a quick in situ overview, he could be second-guessed on his decision-making and communication skills. Congratulations on your diligence, attention to detail, and hard work to get us here. We have had no anomalies on our portion of the trip, so let's keep it that way. You are the most highly trained astronauts in history. My primary objective is to see every one of you return safely with me back to Earth and have the time of your life while your here! Having said that, let me be the first to welcome you to Mars!"

Applause followed, with hoots and yelling.

"Having said that, you all know your jobs, and you understand the risks. Of course, any deviation from protocol must be approved by me. I must reiterate that any unapproved deviation from protocol may lead to disciplinary action up to, and possibly including, confinement to quarters. But of course, you already understand this. It's my job to interpret deviation, so if you have any questions, it must be approved through me. If I don't know, we will seek input from CAPCOM. We're all depending on the individual as much as we are the whole. As in the spokes of a wheel, we all need each other. It is all of our lives at stake. Please continue to be professional and friendly as you all have been so far. I'm already extremely proud of each and every one of you. Stay safe, stay smart, stay alert, and enjoy every incredible minute of this journey. We are the few privileged to have this extraordinary experience. Let's make the most of it."

Commander Carrigan noticed Mary rubbing her forehead. "Does everyone feel okay after exposure to the Martian gravity? Can you all perform your jobs?"

Mary nodded. "Yes, commander."

"Yes, commander." more crew members replied as well.

"If anybody needs more time or would like some down time, you let me know. We do not want anyone getting sick in their Mars suite." Commander Carrigan looked for a response.

"Okay, shift Alpha, we have work to do! Suit up for immediate recon. B Shift, you're on readiness standby. Time is 09:30 Zulu."

There was a lot of commotion as the different shifts readied themselves for the first Mars EVA (extravehicular activity) in human history. Since everyone was already in their Mars suits, preparation for the EVA was greatly simplified.

Deep space can have temperature variations from 1400°F in the sun to a low of around -300°F. Due to storage constraints, a minimum number of deep space suits were brought on Staten, but there were more on *Starships VI* and *VIII*. Mars, with its small atmosphere composed mostly of carbon dioxide, is still better than the moon's, which has no appreciable atmosphere. As a result of this atmosphere, the temperature range on Mars is a bit more moderate, with a high as much as ~70°F to a low of about -120°F.

What this means is that the Mars suits are lighter and more flexible than the bulky deep space suits. Still, any small tear, rip, crack, or penetration would immediately cancel any EVA activity, and emergency procedures would be implemented. Any perforation would mean the immediate application of a special type of reinforced duct tape they all carried in their emergency pack. Each astronaut had drilled extensively in a multitude of scenarios, but the universe is a big place, and no one can predict all possible scenarios or eventualities.

Before *Staten's* arrival, CAPCOM was, of course, aware that an EVA would be conducted as soon as the crew felt strong enough to set foot on the surface of Mars. They were informed that the first-ever EVA on Mars was going ahead as planned. Due to the time lag, all EVA communication and coordination was done with Staten functioning as the command center coordinating all surface activities. Commander Carrigan chose to personally observe Communications Specialist Steven Henderson, who was coordinating the activity.

Steven grew up in Cuba, Missouri, a small town outside St. Louis. When people found out he grew up in Cuba, they would always jokingly say, "How's your Spanish?" He knew it was coming, so he'd always respond with, "No mucho entiendo, pero tratar." If they understood this, they would get a bit of a laugh out of it because it very much destroys any semblance of proper Spanish. If they didn't understand it and thought he spoke Spanish, the last laugh was his, since he knew it was in very poor form. It means "I don't understand much, but I try."

Commander Carrigan liked Steven's midwestern attitude and had a deep respect for his work ethic and intelligent insight. He rarely noticed any work incomplete or overlooked, and if he did, Steven always was ahead of him with answers to his questions or reasons why he was delaying the work. He thought Steven would make a fine commander one day.

5. First EVA on Mars

The first EVA on Mars began one hour after Commander Carrigan announced its commencement. All coms checks completed for each of the seven astronauts, and all equipment verification complete, the equalization room hatch was slowly opened after achieving the same pressure as the Martian atmosphere. *Staten*'s crane was initiated and powered up by engineering, and soon they were standing on the crane deck while being lowered to the Martian surface.

At 10:47 Zulu—or international time—cameras were operating, which meant that the entire world was watching. Even though they had a time delay of nine minutes, it was still considered "live" back on Earth. As the deck bumped to the Martian surface, all eyes were on Andrea Tripp. Mission Specialist James (Jim) Galway unlocked the safety bar and looked at Andrea, who couldn't have been more nervous. Andrea had been elected by her fellow astronauts to take the first step on the Martian surface, and the whole world knew how important this moment was going to be forever in history.

As Andrea's toe caught the edge of the deck, she gasped, "Oh God!" Her heart leaped into her throat as she realized the lip of the deck was slightly higher than she remembered from training. Her recovery was phenomenal, however, as she turned to face the main camera.

She took a deep breath, and in a steady voice, calmly began. "We come as one people, as one race, the human race, for the advancement of all, to be better than we were, to advance imagination and understanding, to make the universe a better place."

As a truly multinational crew, mission specialists Ashanti Sumbika, James Galway, Adnan Ashari, Pedro Lopez, Carl White, and Haratu Suzuki stepped onto Valles Marineris to start the first EVA in the history of Mars, while 760 million people back on Earth watched in disbelief.

MS Andrea Tripp's astounding recovery from "possibly tripping" left people wondering if she had actually tripped because she was so composed with the delivery of her first words on Mars. "Did Andrea Tripp trip?" became the subject of the news and daytime talk shows for the next week.

6. Mars Mission I -
Valles Marineris

Valles Marineris is one of the largest canyon systems in the solar system at more than 2,500 miles in length from east to west, over 120 miles wide, and over 6 miles deep. It was chosen as a landing location for many different reasons, it offered many primary areas of interest, which need to be studied. But mainly because of its low altitude and proximity to the equator of Mars, it offered the warmest year-round climate. The winter will be harsh on this desolate planet so much further away from the sun, so preperations for the approaching severe cold weather were required and programed into the daily activities of the mission.

Priority number one was to secure the foundation of the 180-foot tall *Staten*. The six landing legs of *Staten* were flat on what looked like a dry lake bed. There seemed to be no pitch to the ground at all, which was the best news they could get at the moment.

On further inspection, all aspects of the immediate vicinity matched the mission specifications. There were no soft spots, no holes, no evidence of subsidence or subsurface anomalies such as caves, caverns, or any evidence whatsoever that *Staten* might move. The soundings reflected compacted soil and rock 6 feet down, followed by what appeared to be solid rock around the vicinity. There were no large rocks in the immediate area, either. Temperature readings and humidity levels were noted, atmospheric samples were taken for later evaluation, and even some soil samples were collected to evaluate toxicity and compatibility levels for the humans and plants. There was also no apparent wind.

The astrogeologist and chemist, Barbara Black, was responsible for *Staten's* tiny chemistry laboratory, and she was biting at the bit to get some Martian soil to evaluate.

Many NASA robotic missions sent to Mars have found perchlorates to be a component in the soil. Perchlorates, often a component of rocket fuel, are toxic to humans and most earthbound plants, meaning that the soil on Mars has to be treated to remove the perchlorates prior to using the soil as a growing medium, or they have to find soil that doesn't contain these toxins. Once *Staten's* crew stepped on the surface of Mars, their Martian suit would never be taken back inside beyond the equalization room. They would do a "gross decontamination" cleaning using high-pressure air to blow off their suits and special attention was paid to the boots, prior to entry. The equalization room was vacuumed out after every EVA. Perchlorates are why hydroponic growing systems were identified as a priority for Mars Mission I and testing.

Two 6-foot satellite dishes were laid out as a communication redundancy. The temperature was a balmy 64°F and the humidity 80%, which MS Suzuki thought seemed a little high. The sun was high in the cloudless sky. It really was a beautiful day.

The visibility was remarkable. They could easily see all nine of their starships off to the north, and it was obvious which was *Starship IV*. Even though *Starship IV* was lying on its side, from their point of view, it looked surprisingly intact, as if a giant child had knocked over his toy.

Having secured the immediate surroundings and determined the landing zone for Staten was safe, it was decided to remove *Staten's* rover, or Mars car, and carefully make way for *Starship IV*. The Mars car is somewhat like the Lunar Rover used by the astronauts on the surface of the moon, but with many new toys.

The Mars car had a maximum range of 100 miles with a solar panel roof and required some assembly after arrival. The roof offered protection from solar and deep space cosmic radiation. The Mars car also had clear, flexible vinyl sides that offered further UV protection, and they could be unrolled and zipped into place in case of a dust storm or for protection from severe, cold weather. At the end of each day, the cars were zipped up for protection and plugged in to recharge.

The oversized tires of the Mars car allowed for over 2-foot ground clearance and were over 16 inches wide, allowing traction on the multitude of sand dunes on the surface. They were made of a special rubber polymer to survive the brutal cold and were also UV protected so the sun wouldn't destroy them. The tires arrived empty on Mars and had to be filled with a small compressor that was built into the body of the car, along with a portable atmospheric measuring device (PAM) and a rechargeable flashlight, which recharged using the vehicle's batteries.

The cars also contained an emergency O_2 tank and of course spare duct tape. On the front driver's side, just below the windshield, was a power outlet for portable power tools or auxiliary lighting. *Staten* could track each car up to 40 miles away on a flat surface. Each car also had LED lighting with cameras on all sides and in the rear-view mirror which showed all riders. *Staten* could access these cameras, and if an astronaut was unable to respond or had fallen unconscious, the Mars car could be recalled remotely. Also an injured astronaut could instruct it to return "home," and it would autonomously return to *Staten* on autopilot.

It was time to do some recon for the salvage job they had ahead of them. As soon as *Staten's* rover was lowered and assembled for the quick trip, they were off. It took less than ten minutes to get to where *Starship IV* had toppled over. After securing the perimeter and considering hazards such as leaking fuel, potential hyperbolic explosions, or the possibility that the fallen craft could move or roll, some nearby rocks were used to help stabilize and secure the vessel. Upon declaring it safe and stable, the crew was free to start the assessment.

As a result of her tipping over, *Starship IV* had a crack in her hull, right between the two main fuel tanks which were pressurized methane and liquid oxygen. Her payload appeared to be intact; however, a more thorough evaluation would be needed. MS James Galway took pictures on all sides and included some from a distance to show the entire craft and her position. To off-load her payload would be a lot of work. They needed to spend time on this one.

The EVA had gone on for a long six hours. It was time to return to *Staten* with the results and pictures from their investigation, which were actively being uploaded to *Staten's* computer as they made their way back across the beautiful valley of Valles Marineris.

Commander Carrigan came up on their helmet headsets. "EVA canceled. Return to *Staten,* immediately!"

Lead MS Andrea Tripp responded. "Roger that. Returning to Staten." She stole a glance at Pedro, who looked back at her and shrugged.

7. Sol 1 - No Comms

Once safely inside *Staten,* all-crew was assembled on deck.

Commander Peter Carrigan, stern-faced and obviously upset, faced the crew.

Andrea nervously leaned over to Pedro. "So, what's this all about?"

Pedro looked at Andrea. He was shocked to see her beautiful hazel eyes staring back into his. He stammered, "Uh, the commander is about to tell us."

Commander Carrigan cleared his voice. "Crew, listen up. I have some serious details here to go over with you. Before I forget, I want to give a big thumbs-up to Mission Specialist Tripp. Andrea, good job, and excellent delivery. For the rest of you, I didn't know what she was going to say, but I couldn't have said it better myself."

A nice round of applause followed, and Andrea turned beet red as she nodded to everyone.

As the noise died down, Commander Carrigan's steel-blue eyes took on a look that said pay attention now. "Okay. This is serious." You could've heard a pin drop as he looked around the room. "During the EVA, we seemed to be having communication issues with CAPCOM–Houston. We suddenly had significant static and interference breaking up reception. It was apparent that they were having a hard time hearing us, and we were having a hard time getting clear reception from them. We went from virtually perfect reception to more and more static until we lost them to complete static. We went to the Beam Wave transmitter, which, as you know, is mostly used for data and video, and that isn't working either. I have Communications Specialist Steven Henderson working on it as we speak. We're keeping the working frequency open and checking alternative and emergency frequencies as well. Thank you for being diligent about getting the back-up satellite dishes in place, but they don't seem to be helping. With all of the different frequencies and platforms, even the Deep Space Network seems to be offline. Of course, we will continue working the problem. This was not in any of our playbooks. No policy, no procedure foresaw a complete, extended loss of communication with CAPCOM. We have been offline for well over an hour now, and Steven would have informed me if we had any changes. We cannot foresee all possible outcomes. So, there you have it—an unexpected variable. As far as I see, this doesn't change our mission. We've been operating as an autonomous unit due to the length of time it takes for communications to transmit at this distance. It does not change our mission, but it takes on quite a different perspective. I will notify the entire crew should the situation change. Until then, carry on in the professional manner with which you've been conducting yourself on this mission. Bravo shift will carry on at this time since Alpha shift has concluded."

Andrea looked at Pedro as he was getting ready to return to duties. "So, that's it. We're very much alone, 35 million miles from the Earth."

Pedro looked into her hazel eyes again and felt a familiar pain in his heart. "Yeah, I guess you heard the commander." Pedro was starting to have more feelings for Andrea than ever before. He liked her as a friend and she seemed to like him too. They worked together in training like best of friends often interpreting a look or an intonation before anyone else in the crew.

Andrea started mumbling. "No email, no communication with our families. That's going to be hard on them."

"Yeah, I guess so, but let's hope they get whatever the problem is fixed, because it doesn't appear to be on our end. We are getting active responses from the reconnaissance orbiter and our communications satellites, so it's not us. The problem is with Earth."

"How could Earth be the problem?" Andrea looked away. "That doesn't make any sense."

8. A . . . What?

According to the Fort Lauderdale Airport radar, the anti-satellite weapon known as an ASAT exited the water about a mile off Bob and Brian's 27-foot World Cat port bow at approximately 70 mph. By the time the Miami International Airport radar picked it up, it was approximating 160 mph. Thirty seconds later, Homestead Airforce Base spotted the "anomaly" exceeding 250 mph.

Key West Naval Air Station radar, having been alerted by Homestead Airforce Base, visualized it on radar as it "broke the horizon" approaching 500 mph. Raymond Juarez, lieutenant first class, was a new radar technician fresh out of radar school. He was on his first assignment, and one of the few radar technicians on the base, qualified to operate the new HP interferometer and infrared-equipped radar. Raymond watched as his system visualized the anomaly and split it into two images. The heat refractions indicated an increasing amount of heat, speed, and altitude.

Colonel Barry Diggs was notified that some kind of bird was flying at breakneck speed in his territory without known authorization. And he wasn't happy. Heads were going to roll, and someone was going to lose their new rocket.

He commanded that an all-ships-all-aircraft alert be sent out immediately to anything within 300 miles with corresponding latitude and longitude. Imaging was badly needed. Could he get a satellite on this thing? He ordered his support personnel to contact NORAD. This was not going to be an easy day, and after his previous night, he didn't need this BS.

Colonel Diggs arrived in the radar center just as Lt. Juarez was calling out the numbers, calmly registering the target's speed. "Approaching six hundred miles per hour."

Colonel Barry Diggs demanded, "What is the location?"

Everyone in the room turned toward the red-faced colonel, including Raymond, who quietly said, "Up. It's going straight up like a frickin' moon launch . . . uh . . . sir."

Both F-16s standing by in preflight readiness mode were launched before Colonel Barry Diggs had been notified. This was done just in case this was a real target, they could always be recalled. They were removed from preflight readiness mode to active status within ten seconds of notification, airborne after sixty seconds, and achieving 500 mph with afterburner engaged four minutes after the alarm.

They had the "bogey" on both of their radars, but it looked weird. It was going straight up!

Tower control came on the radio. "This is an unauthorized missile launch of unknown origin obtaining status as we speak. What is your range to target?"

Captain Robert Roberts, call sign Zulu (because he was always the last one to roll call), responded, "We are approximately five hundred miles and closing."

Both F-16s broke Mach 1 at 742 mph.

Tower Control replied, "I show you at 30,000 feet and climbing."

"30,923 feet and climbing," Major Tobias said.

"What is your range?"

"Not good. Bogey is approaching Mach 2 and continuing to accelerate. Will pursue."

Both F-16s were completely vertical, trying to chase what looked like a rocket launched out of Cape Canaveral, but they weren't going to catch it.

The control tower hailed Captain Roberts again. "Zulu, tower control."

"Go, tower. This is Zulu," Captain Robert Roberts reported. "Approaching 50,000 feet. Return to base required prior to loss of apparatus."

They were out of gas. They had burned 10,000 pounds of jet fuel in fifteen minutes and never even saw their target.

"Loss of apparatus is not an option return immediately," was the response from the control tower.

They weren't allowed to ditch their F-16s in the Gulf Stream, and they had to return before running out of jet fuel.

Colonel Barry Diggs couldn't believe what he had heard and seen in the radar room, and as he headed back to his office, he also couldn't believe what he had to do. *"Oh, shit this is going to create a ton of paper work."* he thought. He had to make a phone call to Southern Command.

Admiral Thomas Fanning, commander of Southern Command, was sitting in his office in Doral, Florida, reading emails and trying to get caught up on his generous list of responsibilities. He had come in early before the start of another hectic day. At 08:30, his phone rang, and he doubted his secretary was at her desk since it was café-con-leche time.

On the fourth ring, he picked up, sorry to give up the headway he was making with his emails. "Admiral Thomas Fanning here," he said, revealing a little more of his frustration than he would've liked.

"Admiral Fanning, this is Colonel Barry Diggs, Key West. We have a situation here . . ."

As soon as Admiral Fanning hung up the phone, he punched up the secure line for the Pentagon. His day had just taken on a whole new meaning.

"Joint Chiefs, this is General Taylor."

Admiral Fanning spoke without his usual greetings and chitchat. "Eugene, this is Tommy at Southern Command."

General Taylor knew by the tone of Tommy's voice that this was a serious call. He drew in a deep breath as he got ready for the bad news. "Yes, Tommy."

"Key West and Homestead scrambled a total of six F-16s less than thirty minutes ago to investigate an anomaly launched twenty miles off Fort Lauderdale. It appears to have been directed straight up into space. We've been in touch with NASA, and they knew nothing about it. NORAD has been alerted. It was faster than our F-16s and went off our radar in excess of Mach 2. I'm routing the electronic surveillance data, including corresponding radar images to you immediately. We have a Coast Guard Cutter heading into the launch vicinity, ETA ten minutes. I will copy you as soon as we have more."

"I'm convening the Joint Chiefs immediately. Who knows about this?"

"Only me at South Com, probably most of Key West, Joint Base by now, most of Homestead Airforce Base, and as far as I know, MIA and Fort Lauderdale radar operators tracked the target."

"Okay," General Taylor said, "send investigators to Fort Lauderdale and MIA radar operators under strict orders to investigate but not talk about the incident. Repeat, they are to reveal nothing. And make sure Key West and Homestead keep a lid on it until we know more. Update me the minute you know something. Anything further?"

"No, sir."

9. Now You Seem 'Em,

Now You . . .

KH-11 Kennan, one of the largest satellites ever put into orbit, run by the United States Department of Defense–NRO (National Reconnaissance Office), weighing in at 43,211 pounds, was no longer in existence after 12:12 Zulu. It experienced a rapid, unexpected disassembly due to a high-speed collision with what could only be described as a missile that had just left Bob and Brian's favorite fishing spot. Now two scared and confused, retired firefighters, one with a bandage wrapped around his head, were on board a United States Coast Guard Cutter, giving their statements with their fishing boat in tow. Fishing was canceled for the rest of the day.

At 12:18 Zulu, Terre Star 1, one of the largest communications satellites ever sent into orbit, experienced the same fate as KH-11 Kennan. Both water-to-air ASAT missiles triggered similar responses from authorities, but virtually nothing could be done about either one. Neither rocket seemed to pose a threat to any territory of the United States or its territories. NORAD, Southern Command, nor Colonel Barry Diggs understood the implications of the rocket launch, not that there was anything they could do about them. The only thing the tracking stations could do was watch. No one understood what they were seeing or what was to follow.

Satellites are made of many different kinds of materials such as aluminum, titanium, plastics, silicone, glass, different epoxies, and exotic metals. Terre Star 1 had its own propulsion system and fuel. The International Space Station (ISS) has been damaged from debris smaller than a millimeter in diameter. Paint chips damaged the shuttle windows due to collision at speeds exceeding 12,000 mph. Collision impact forces are exponentially increased with speed. The minimum speed required to maintain orbit and not fall back to Earth is 17,000 miles per hour, which is faster than a bullet fired from a gun. Usually, the higher the orbit, the higher the speed required, especially if it's a geostationary orbit such as the KH-11 Kennan.

The United States Airforce tracks all space debris over 4 inches via the United States Space Surveillance Network (USSSN). There is over 1 million pieces of debris smaller than 1 centimeter and over 900,000 pieces of debris between 1 centimeter and 10 centimeters, and 34,000 pieces of debris over 10 centimeters thought to be space junk currently in orbit. That was before the two satellites experienced their destruction, along with the speeding rockets that hit them. The resulting impacts created millions of new pieces of debris heading off into all kinds of incalculable directions. The knock-on effect was insult to injury: more collisions with more satellites creating more debris, which caused more collisions with more satellites.

This exact situation was postulated by NASA scientist Donald Kessler in 1978 and has since been labeled the Kessler syndrome—or Kessler Effect. Within three minutes of the first collision, the alarms began sounding on the ISS, which had happened before. However, this time they were informed by Houston's Flight Control to batten down the hatches and immediately prepare for emergency evacuation.

This has never happened before! There was no time for the usual fully planned, organized, and coordinated orbital boost conducted by Houston. They had to get out of Dodge now.

As Commander Amhurst did his best with the emergency move to a higher predetermined orbit, the rest of the crews began emergency evacuation procedures. Since the largest pre-existing debris fields are in low earth orbit, it was thought that the best way to avoid collisions was to move to a higher orbit. The astronauts in the ISS knew exactly what this meant; it was one of their worst nightmares come to life. They would immediately evacuate the ISS one compartment at a time and seal the hatch of each compartment as it was evacuated. Then, as fast as they could, they would get into their space suits, hunker down in the two Soyuz capsules, and hope for the best.

Typically, the ISS is in orbit around 250 miles above the Earth and weighs in excess of 100,000 pounds. Today, they were going to try to move that mass 250 miles higher in a virtually uncontrolled ascent. Who knew where they would end up, but 500 miles in low earth orbit was the target, and they didn't have much time. Little did anyone know this new orbit would be of little help with what was heading their direction.

10. United States Space Surveillance Network

Only one hour to go, and Lieutenant First Class Natasha Brent, USAF, was looking forward to going home and getting some sleep. She was new to this shift work stuff, and so far, it wasn't sitting with her too well. C shift was slowly starting to drift in for the 0600 shift change, and she couldn't wait.

She had been in the Air Force for six years now, and just a couple of months ago, she eagerly accepted a position at the USSSN. She had been in training for her new position for six weeks and had only been at the computer console for just over one week. As far as she could tell, the job was pretty boring, but the night shift was tough.

Natasha glanced up at Amanda, who was coming in to replace Todd. "Hi, Amanda. Good morning."

"Morning, Natasha. How was your shift?"

"You know—same old, same old. Wait. . . . What was that? Something odd is happening. One of my satellites just like. . . . Where did it go? It just disappeared. Oh, my God. It's all over my screen! It's in pieces!" Natasha did something she never thought she would do in the quiet, peaceful computer monitoring surveillance room. She screamed.

The USSSN was quickly overwhelmed. The destruction of KH-11 Kennan alone was more new data than they could handle. When Terre Star 1 was destroyed, it was impossible to track everything. So instead of tracking individual pieces, they began tracking the debris fields as they spread out over thousands of miles at first, and then millions of miles in 360-degree space. The USSSN saw more debris as additional satellites collided with the remains from previous collisions. Some of the satellites were GPS navigation satellites, some were communication satellites, some were private, some were French, Japanese, Chinese, Italian, Brazilian, and Russian, and they all suffered the same fate . . . and created more debris.

As the debris fields grew, something started happening to the radar screens, worrying the technicians. Each of their screens became fuzzy and started to show less of the debris until all of their screens slowly went white.

Engineering was called to fix the problem, after they went through everything and reported back, the technicians learned it wasn't their equipment. Different radars were getting the same thing. The other radar sites reported back on their secure landlines, and yes, they too were getting white-out screens with the worst interference they had ever seen, and there was nothing they could do about it. Worse yet, not even the engineers knew why.

"The radar screens are going white with some kind of massive return," one of the engineers quietly said to his partner.

What they didn't know or couldn't possibly know was that the debris fields were creating a giant static charge on their own. They began interviewing the radar operators and monitors about the last thing they had seen. The engineers listened quietly as the technicians described the disappearance of the satellites, the creation of the debris fields, and how their screens slowly went white. One by one, the engineering teams filed out of the satellite surveillance room and into the adjoining conference room, shutting the door behind them.

They knew what it was. The aluminum, silicone, plastic, rubber, titanium, exotic metals, and everything else satellites are made of, were mixing and blending and rubbing and bumping each other at thousands of miles per hour, causing a static charge with multiple strengths and intensities. The debris and its static charge were acting like a curtain, quickly wrapping around the Earth and absorbing any radio waves that tried to penetrate it, thereby cutting Earth off from any incoming electronic communication or signal trying to leave Earth.

"It's not that we aren't seeing anything," one radar technician noted, "it's that we're seeing everything. It's like the Earth is enclosed in a giant Faraday cage."

No satellite communication would be possible until it was cleared or at least reduced. All space operations were effectively on their own, with no ability to communicate with anyone on the ground, and no one knew how to stop it.

The president of the United States made an emergency address to all Americans on all television and radio stations and social media.

A somber President looked into the television camera. "I just left the Joint Chiefs. We were meeting regarding what I am reporting to you now. We are in a de facto state of war. We just don't know with whom, as of yet. Two major American satellites, one belonging to the United States Department of Defense—which served 'untold' American interests far and wide—and an American communications satellite were both just d—"

All television screens went white with static. All commercial airborne aircraft were contacted via their VHF radios to make it to the nearest appropriate airport and land.

All commercial aircraft were temporarily grounded. Pilots following their GPS guidance systems suddenly started receiving 'gibberish'. They were useless. The GPS system was beginning to fail. Air traffic controllers were immediately overwhelmed with thousands of flights trying to land all at the same time. Until further notice, air travel would be "emergency" only.

Fortunately, local radar still worked just fine. Hundreds of thousands of people were marooned at airports all across the country, many of them with no way to get home. Rental cars started selling out quickly. Cell phones wouldn't maintain connections and eventually became unusable. And this was only the beginning.

The United States Coast Guard began emergency radio transmissions to all ships at sea. They were told via their VHF radios to change to paper charts if they were within radio range. The fax machines on these behemoths of the sea went quiet, and communications with their headquarters were lost. Radio was all they had to communicate since cell phones were down, satellite phones weren't working, and VHF is only good for "line of sight."

All shipping ports prohibited departure of any ships that weren't already docked, since GPS navigation was unusable. The vessels in port were not allowed to leave port until further notice. The internet crashed, and banking transactions started to slow to a crawl as the cell phone network continued to fail. Trading of any kind stopped overnight. Most purchase orders were sent and received via the internet, so no trading of any kind was able to continue. The stock market went into an immediate headlong dive with panic selling, and then it went quiet as selling slowed to a crawl, then stopped completely, without the internet. Retail prices started to skyrocket when people started panic buying. The grocery stores in all major US cities and most of Europe emptied out as if a hurricane was imminent.

The ATM machines stopped working, and credit cards were worthless, making cash the only means for making purchases for food and gasoline. All trains across the country were forced to a crawl, their computer-operated control systems crashing due to dependency on the internet. All train conductors were forced to reduce speed to 10 miles per hour, and once they attained the nearest terminal, they were forced by safety officials to stay there. No planes and no trains.

11. This is going to Get Interesting . .

While everyone was focused on the communication issue, Commander Carrigan was busy doing his rounds, checking in with each of his officers, and ßaking sure no other changes had occurred. He explained that their logs needed to be complete and more detailed than ever since CAPCOM–Houston wasn't getting daily updates, and their automated systems were no longer communicating.

What the devil is going on with Earth? he thought. *Now we really are alone. This is going to get interesting.*

He considered how his crew, no longer able to communicate with family or friends would be affected. This was the most professional crew he had ever seen assembled together in one place, but still, the loss of all comms hadn't been considered any further than the loss of an individual radio or some system failure. He knew the problem was back on Earth.

Twenty-seven-year-old Andrea Tripp was having a sudden surge in popularity amongst her peers. Some people would look at her funny and walk away giggling. Some people wanted to know what it was like to be the first to walk on the surface of Mars. She would just laugh and say, "I'm glad it's over." Others wanted to know what the Martian surface was like. "It's kinda crunchy." she reported

Everybody's "walk" on *Staten* was a little weird due to the new low gravity, which would take time to get used to. Since the gravity on Mars is approximately one-third that of the Earth, astronauts tend to hop a little. The whole crew was trying to adjust in different ways. Some did a kind of shuffle, some did a scoot, and it was rather humorous. It made the 360-degree jogging path designed for use in zero gravity very difficult and interesting—if they were courageous enough to try it.

Even though communications with Earth were down, everyone on *Staten* was in a really good mood now that they had made it safely to Mars. Their dreams had come true. And they were all sure that the communication thing was going to get sorted out sooner or later. Even CAPCOM couldn't do anything to help them if they needed it immediately, so what difference was it in the short run?

Day two on Mars was a full-blown schedule. Commander Carrigan had reorganized the shifts a bit, rotating some people around and reducing C shift in personnel since more crew were needed during the EVAs, and there were fewer positions requiring attention at night now that they had safely landed on Mars. Seven astronauts were staged for ready deployment, while seven astronauts were sent out on an EVA to verify the effectiveness and productivity of *Starship I* and *Starship II*.

They needed to off-load another Mars rover and its trailer from *Staten* for moving equipment around. They needed to revisit *Starship IV* and see if they could get the hatch open on the far side and get some pictures of the condition of the supplies inside. They also needed to set up the solar voltaic cells with the stands to start charging the battery-powered equipment, including *Staten*.

Commander Carrigan was always around doing something or checking on something. He could be found moving about *Staten* even in the wee hours of the morning, and then disappearing in some hatch, checking inventories, fuel pressure, water, food, life support, or atmospheric measurements. It was amazing how he seemed to be omni present at all hours of the day. Or he might be found sitting in comms with Steven or working with Carl, the engineer of all engineers.

Carl knew everything about hardware, metals, components, welding, tools, and forces. You name it, Carl was your man. Commander Carrigan continued to consider their situation from every angle, constantly apprising himself with the situation. He never wanted to be surprised by a situation such as the loss of all communication with Earth. Occasionally things simply could not be predicted.

Never before in history had men and women ever been so far from the place they called home, literally at work twenty-four hours a day. Perhaps the early seagoing vessels were away from home for years at a time, the conditions back then were close to slave labor. That was a different time.

No one in his crew was married—at least not legally. It had been decided that three years separated from a spouse would be devastating to the emotional bonds of a permanent relationship. They may have a long-term partner, but that was up to them. NASA always had a no-comment status on sex. If it was brought up, all they would say was that all astronauts are professionals and adults. That is it. It was also an aspect of their training that humans have a strange tendency to subgroups within the team, which can cause division. That's why there was no permanent A, B, or C shifts. That's also why he constantly rotated people through different positions, in an effort to limit people creating "territories," or assuming command of public spaces and defining a workspace as "theirs."

Being on Mars made the shift rotation a little easier, but it also made it a little harder. Only Carl and the comms team were rarely rotated, and that was a problem he had to consider.

I think I need to talk to Carl and Steven, he thought. *It's time to send them out on an EVA as well. And I guess I should schedule myself into the EVA roster.*

Starships I and *II* were a mile off to the northeast where the current EVA team was reporting that the water, O₂, hydrogen, and methane were, in fact, being produced, as NASA had reported prior to their departure. This was great news. The methane rate of production was within the parameters of expectation. Excellent news.

That was one big worry off of Commander Carrigan's mind. The autonomic systems producing the O₂ and hydrogen obtain water from the atmosphere. When the sun warms Mars over 32°F, the water in the soil melts and converts almost immediately to humidity or gas, which the systems on *Starship I* and *II* condense into drinkable water. When the temperature-controlled water tanks are close to full, they start to break it down, via electrolysis, into hydrogen and oxygen in separate but attached equipment. They're triggered to do this by a water-activated switch near the top of the tank, so the tank always stays nearly full. This would act as a primary water source for *Starship* X's water supply system when it runs low. However, this would only work during the Martian "summer," when the atmospheric temperature usually rises above 32°F during the day in Valles Marineris.

During the Martian winter, the temperature will rarely rise above 32°F, even in Valles Marineris on the warmest of winter days. Even the autonomously heated water in the water tanks will freeze in the water lines before reaching *Staten*. *Staten's* water tanks held enough water to maintain twenty-one astronauts for about six months. It was understood that for at least three months of the year, they would be completely cut off from their water source, but first, they needed to find and install the water lines. The water lines were listed on the inventory schematic. But they were about a mile away from *Starship I* and *II* and for some unknown reason NASA had packed the water lines in different starships.

Each starship was 'normally' in communication with a computer at MMI command center inside Johnson Space Center, but they were also designed to work autonomously, for a period of time, in case of a communication disruption like what they were experiencing now.

Commander Carrigan was in his office reviewing the operations procedures for starships I and II. *How long would they stay active on their own before requiring a signal or command?* He was hoping indefinitely, but he didn't want to find out after they shut down. Could they be started back up if they shut down? He was pretty sure Carl could do it with some help. Still, he had thousands of pages to search through just on *Starship I*. He wasn't going to leave the lives of his crew open to chance.

12. *Starship IV* - Mars Sol 3

Every Starship from I to X had parts, tools, and redundancies on board to protect MMI from failure, including multiple redundancies for communications. Commander Carrigan instructed Communication Specialist Steven Henderson to evaluate the other remaining starships to see if they were communicating with Earth or each other—or anything. The answer was no. They were actively pinging but receiving no response.

"If they are actively pinging, are they asking for something? Some direction? Or are they just saying I'm here?" Commander Carrigan said.

Steven replied, "The latter. That's what they do when they don't receive any guidance twenty-four hours after landing. It's like Big Ben, but instead of every hour, it's every twenty-four. It doesn't take much battery to send out a ping once a day, and if someone is listening, it'll help them focus their dishes or modulate the frequency if needed. It's also pretty convenient to know when they're going to ping, so you know exactly when to listen for it instead of listening for twenty-four hours a day to pick it up."

Commander Carrigan thought to himself, tapping his temple with this index finger, and then looked at Steven. "Will they ever shut down if they don't eventually get a response?"

Steven frowned a little. "Hmmm . . . let's see. My initial answer to your question would've been no, they won't ever shut down. Then, you know, that word ever is a very long time. Let's change the time frame and think about what the computer programmers were thinking with the design. The starships were designed to always be in communication with MMI Houston. If I were designing the computers for the starships, what logic would I follow if the starships cannot or do not get a reply from MMI Command? You know what I would do, Commander? I would program the starships to continue with their daily duties and actively ping as scheduled unless the batteries go below twenty percent."

Commander Carrigan looked up at the ceiling, absently rubbing his five o'clock beard. "Why twenty percent? Why not ten?"

"If the batteries go too low," Steven said, "you could threaten some of the life-support systems needed to keep the computers and other equipment alive and functioning. If they power all the way down and completely discharge, then some of the more sophisticated and sensitive systems that need to be kept warm would be subject to the ambient temperature. Normally, that isn't a problem, but during the winter months or during a dust storm, the deep cold of, let's say a hundred degrees below zero Fahrenheit, may crack or break anything that has even the smallest moisture, condensation, or liquid cooling systems. And it would never work again. And some of the systems require a bit of power to restart. If the power gets too low, booting them back up may be a bit tricky if not impossible."

Commander Carrigan stood and paced slowly behind his desk. "How long do you think it would take for the batteries to drain to twenty percent?"

Steven had to think about that one. "I guess that would depend on what functions they're programmed to perform autonomously and the weather."

"Okay, so different starships will drop to twenty percent at different times depending on the power drain?"

"Yes, I would think so, but I'm pretty sure we're talking years here, not months."

Concern fell over Commander Carrigan's face. "Are you sure about that?"

"Well, no, I didn't program them, and I wasn't involved in any aspect of systems designs, nor were any of us trained on what to do if we or any of the starships lost long-term communication with MMI Command. It was never considered. But that is how I would've done it."

"Is there any way to know which starships will lose power first?"

"Well, we could send Carl out to evaluate battery strength on one of the next EVAs."

Commander Carrigan continued now pacing behind his desk for a few seconds, then sat down again. "Could we answer their ping? You know . . . figure out what frequency they're listening to, and tell them—"

"No, we can't," Steven said. "Well, yes, we could, but they'll be looking for the correct authorization to start responding, and who knows what they might do without the correct authentication sequence. Probably nothing but, I'm pretty sure each one has a separate authorization code or administrative access code and frequency."

"You were reading my mind."

Steven smiled quizzically. "Of course, they would be protected from some random remote access code, but perhaps we could do something manually. The computer programmers wouldn't expect anyone to just walk up to them on Mars, open the door, and start trying to fire up the operational systems."

"Access code?"

Steven shrugged. "Probably. Every computer guy locks up his stuff."

With a heavy sigh, Commander Carrigan admitted, "Yes, I'm sure you're right."

13. *Starship IV* - Mars Sol 4

The Mars Reconnaissance Orbiter (MRO) is still an operational satellite circling Mars, launched in 2005. Built by Lockheed Martin in coordination with NASA's Jet Propulsion Laboratory, it is truly a testament of durability and the dedication and hard work of the engineers of both of these operations.

The MRO constantly circles Mars, taking pictures with different high-resolution cameras, spectrometers, and radar, providing excellent detail for weather data, mapping, and potential landing locations. The MRO produced beautiful pictures of *Staten* and all nine of the other orbiters, which Commander Carrigan and his teams used as a detailed map to plan their EVAs.

The landings were remarkably close to the planned layout by NASA. The third EVA was planned and organized. B shift suited up. C shift prepared as ready backup with two personnel from A shift. C shift had been reduced to a crew of only four personnel since they had previously been the midnight shift until today.

The Mars car and its accompanying trailer were used to haul equipment to the fallen *Starship IV*. It had been determined that *Starship IV* had fallen over after "safely" landing with two of its six legs in a slight depression. If one leg had landed in the depression, it wouldn't have fallen over, but two was enough to start the momentum movement of the 180-foot rocket as it settled and slowly rolled onto its side.

The onboard radar-lidar system should have recognized the variance before it authorized the final landing sequence. It was easy to see while standing next to its final resting position that from a visual perspective, the ground looked flat. It wasn't until they walked around the landing zone that they saw a gradual hill consisting of mostly sand and dust.

MS Ashanti Sumbika realized that the landing engines might have created the depression *Starship IV* fell into. As the thrusters blew away the lighter sand on one side, it probably created the uneven surface, which explained why the landing sensors didn't pick it up before landing: they couldn't "see" through all of the dust and debris that the thrusters created upon landing.

Ashanti explained this to Steven over her helmet mic, and she took pictures from her helmet cam, so they could be entered into the log along with a report of the incident for

MMI Command. This situation probably hadn't been considered while designing the starships.

There were three primary designs for the starships on this mission. *Staten* was the only crewed starship. *Starships I* and *II* were the tanker starships that produced in-situ vital resources and also carried a small amount of cargo. The supply starships *III* through *IX*, carried the bulk of the vital supplies and equipment, which rounded out the three main design platforms.

Carl White, acting lead for today's EVA, reevaluated the safety of attempting to make entry into *Starship IV*, and after a few minutes, decided that the enormous starship was stable, secure, and would not shift or roll. All of the high-pressure fuel and propulsion tanks appeared to be intact, and their integrity was not at risk. At least not at risk right now. Things change.

The EVA crew set up the ladder they had brought with them from *Staten*. After making sure the ladder was safe, Carl climbed up and attempted to open the hatch thirty feet off the ground as two crew members held the ladder to stabilize it. Carl rotated the manual hatch with a tool that he had brought with him from *Staten* and popped it open. It looked pretty good inside since everything was shrink-wrapped and thoroughly strapped in using nylon netting attached to aluminum framing for the violent launch and subsequent landing. All of the stored items would have to be manually carried off and there was a lot of supplies.

Thank God for the lower gravity on Mars, Carl thought. I don't see anything permanently damaged other than the ship itself. It suddenly occurred to him, *how the hell are we going to get into the other starships? Starships I and II were activated two years before, so they're up and running. But have the others been turned on for astronaut use from MMI Command? It is not like we can click on a manual switch or access panel at ground level, and I've never heard of any kind of remote access. MMI was supposed to activate them all upon Staten's arrival, but did they have time before communication was lost? Oh, this could get bad.*

He off-loaded another set of solar panels and passed them down the ladder to Haratu.

The more Carl thought about it, the more he didn't like it. *Was it possible that they had everything they needed but couldn't gain access? The access hatches were 150 feet in the air, and even after the first stage boosters returned autonomously to their predesignated landing site, the starships were still over 180 feet. They didn't have any ladders that would go that high, and a fall from that height would probably be fatal, even in Mars's reduced gravity.*

I have to talk to Commander Carrigan, he thought. *I'm sure, he's aware of this situation, but he hasn't spoken about it with me or the crew yet. There's probably a good reason why he hasn't.*

Carl calculated how long they could last on Mars without any more supplies than were on *Staten* and starships *I* and *II* and whatever they could carry off the stricken *Starship IV*. He knew that starships *I* and *II* didn't carry much in the way of supplies since they were tankers. As he calculated, he started to get nervous.

Meanwhile, Commander Carrigan was in his study with Steven, who had been relieved by Chief Pilot Mary Pfeiffer.

After listening to the telemetry replay off of the automated recording on the *Staten's* computer, he said, "So far, all we know is CAPCOM seemed a little preoccupied in the last three minutes, by my estimation, prior to loss of signal. This is unusual for them. When we received the final transmission that said, 'Communications may be compromised."

Steven nodded in agreement. "Yes, that is unusual. I agree. And there's nothing further to indicate any issues after that. No explanation, nothing."

"No, I've been over them at least ten times trying to eke out some more information, but that's all we have to go on. Frustrating!"

"So, what do you think happened?"

Steven sat back in his chair, absently scratching his head. "What do I think?"

"Yes. What do you think?" Commander Carrigan said. "Your best guess."

Steven had been a little nervous during this chat in Commander Carrigan's study, but now he was confused. "You want me to guess?"

Steven had never guessed on anything related to the mission, and he was pretty sure Commander Carrigan hadn't either. They always dealt in facts, not fiction. With frustration, Commander Carrigan said, "I've been thinking about it, and since we don't have any details, I need your best guess. I depend on your insight at times like this Steven."

Commander Carrigan took on a sober look. "Look, it's going on three days since we lost communication. I know you and the crew have been thinking about this, too. What on earth could've happened to lose communication for almost three days? You know it couldn't have been a power outage or some strange power surge. Besides that, the Beam Wave system is down, too. You know as well as I do that it isn't just CAPCOM–Houston or one of the three Deep Space Network systems in three different countries all around the Earth, which all have automated back-up-power-generating abilities and nine different transmission and receiving dishes on multiple frequencies. So, this situation simply should not happen. Yet, here we are. So, what do you imagine is going on—or could have happened—back on Earth?"

"Well, you're right. I have been thinking about it, and of course I'm not the only one on the crew wondering about it. Everyone has their thoughts or ideas—a nuclear attack or a particularly strong solar flare or a CME. I think those ideas are both wrong. We would've been told via at least one of the Deep Space Network dishes that something like that was happening, and the Space Weather Network would've had plenty of time to warn us of a strong solar flare. So, I was wondering about an electromagnetic pulse attack. But again, that wouldn't take out all of the communication systems around the world."

"So, where does that leave us?"

"We're sticking to protocol," Steven said, "attempting to make contact with CAPCOM on all frequencies, every hour on the hour, twenty-four seven. Maybe they can hear us and we just can't hear them."

Steven sat through the silence as he watched Commander Carrigan think it through.

"Obviously," he continued, "the other starships aren't receiving any data or signal, either, or they would've been taken out of hibernation mode. We've put up three more dishes for improved signal strength, and we're able to communicate no problem with the MRO. No, CAPCOM knew something was up on Earth, or they wouldn't have reported possibly compromised communications. Even they knew it wasn't us. They could see something was happening—they just didn't have time to explain it. How is this even possible?"

"If you come up with any more ideas on the communication issues or concerning the other starships' situation, please let me know. I need to bring up these issues with the crew soon. It seems ludicrous that all the supplies we need are here, less than one mile away, yet we have no way of getting to them," Commander Carrigan regretted saying.

"Yes, sir. I'll start going over the MMI procedures and see if I can find a back door or alternative means of entry into starships III, and V through IX. We need those supplies pretty soon."

"Yes, we do," Commander Carrigan said, "and thank you, Steven. I don't think you realize how much I value your feedback."

Steven was surprised to hear the commander say this in person. He usually gave feedback in group settings, and this felt personal and gratifying. It also spoke volumes about their predicament. Previously, he hadn't really considered how dire their situation may actually be.

"You're welcome," was all Steven could say, which sounded a little shallow to him, but now he felt more motivated to search the voluminous procedures manual, which could take days, even with the computer's help.

14. Johnson Space Center -

Hell Breaks Loose

Brad Brown had seen a lot in his twenty-four years at NASA and over the past four years at Johnson Space Center. But the past three days had been beyond belief. As he faced his fourth twenty-four-hour shift in a row, he decided if he ever survived this, he would write a book.

Yes, he had taken three, three-hour sleep shifts in the breakroom, but he never left the building. And even though he needed the three-hour breaks, on the first two, he didn't get any sleep at all. The third night, he was pretty sure he slept out of sheer exhaustion.

He was still trying to get his head around the last three days. Brad had been interrogated, briefed, addressed, re-briefed and debriefed. He had been in meetings upon meetings with his boss, Gerald Adams, the director of NASA. All Brad understood was that some rocket blew up two satellites, neither of which belonged to NASA. He knew this was no accident. No one told him, but these kinds of things just didn't happen twice in the same day, at almost exactly the same time.

Someone, **he thought,** or *probably some country. . . . Is it possible that a single individual would or even could do this?*

The more he thought about it, the more he realized that the technology had definitely been dribbling down to surprisingly pedestrian levels.

Or are we doing a better job at education?

Brad wasn't an investigator, but the net result was the full Kessler Effect—something he had dreaded since he heard about it in college at MIT. At first, everybody poo-poohed it as totally fictional. Unfortunately, it was real, and it did happen. He was hoping that it wouldn't happen until after he retired, if ever, but no such luck. The theory or hypothesis was that if this ever occurred, space would be closed for business for years to come: No NASA, no ESA, no satellites, no space station, no lunar landings, no Mars... Nothing for years, maybe decades, unless someone could figure out how to clean it up.

There were twenty-one astronauts on Mars and six on the ISS, if it was still in one piece. What would that mean for them? No one knew.

Brad had worked his way up from a simple entry-level technician position in NASA to his present-day position of CAPCOM and one of the mission directors for Mars

Mission I. Just four days ago the final starship had landed safely on Mars, right on target. They had had one major anomaly when one of the starships had fallen over, but at least it didn't explode. Of course, the atmosphere on Mars is mostly carbon dioxide, which helps reduce the possibility of an explosion occurring. From the pictures he had seen from the MRO, it looked to be salvageable. With all the redundancies built into the mission, they were easily still a go for the mission, with one big exception: the astronauts on Mars couldn't get to their supplies. Of course, they couldn't turn around and come home now. Not today. Not for another year. And then they had a six-month return journey back to Earth.

The first EVA on Mars was a total success. As the whole world watched, NASA was once again experiencing a huge popularity revival. Shortly thereafter, CAPCOM and all space-related communications began experiencing static interference. It slowly worsened until all of their telemetries became worthless. All radio comms were down, and even the Beam Wave technology and the Deep Space Network were useless. They were effectively cut off from the twenty-one astronauts involved in the first manned mission to Mars.

The timing of the communication disruption occurred just as the procedural sequence was initiating to instruct the remaining six starships to awake from hibernation and begin their startup process—an order that could only be given from MMI Command.

What a design flaw this is, Brad thought. *We lost communication just before we were able to give the command via the Deep Space Network.* Soon after loss of communications, they were informed of what had happened. After all of the briefings and meetings, there was nothing MMI Command, CAPCOM, or any of the mission directors could do about it.

Brad was deeply worried about his crew. It was almost like they'd sent them on a camping trip and locked up all of their supplies. This was turning into a human survival trip—not a Mars exploration mission.

We need to send that command!

Brad felt sick, as though they had abandoned the whole crew on Mars, and he had a deep foreboding for them.

Outside the walls of the Johnson Space Center, the situation was beyond desperate. Almost all of the developed countries had become so dependent on satellites for day-to-day activities, and no one stopped to realize what would happen if they lost them. It wasn't only the satellites for use at NASA, and not only in the United States, but all of Europe, much of Asia, Canada, Mexico, and most countries in South America. Hell, even Russia was dependent on satellites.

After what he had seen and heard was going on outside of the walls at Johnson Space Center, it looked to him like a third-world nation was a pretty good place to live right then. His wife and kids reported to him on the landline that conditions around Houston were as bad as he had feared and were probably going to get worse soon. Fortunately, Brad was at his desk when his wife found a landline after the cell phone system went down. He told his wife to take their two kids, all of the cash and canned food and water at the house, and head out to her dad's farm 40 miles outside of Houston. He hoped that was far enough out of town before the real craziness set in. Her father had been a military policeman in the Army in the 1990's and early 2000s and had done a few tours in Afghanistan. He would be able to take care of them.

Brad had heard that other than emergency flights, all domestic and international flights had been grounded. The national defenses were on high alert. F-16s and other military aircraft could be seen and heard flying overhead. The National Guard had been called in to help reduce hysteria and limit the looting. Any kind of banking was impossible since banks relied on either the internet or wireless computer systems. The federal government had imposed a national price freeze after there had been runs on grocery stores and gas stations and pretty much everything else. Not that a price freeze was going to do any good. Grocery stores could only take cash, and their shelves were already empty anyway.

Anyone without an old-fashioned, television-style antenna or rabbit ears wasn't picking up TV channels. Fortunately, the radio worked for the stations that were still operating their systems in the traditional sense, but there was no national broadcasting of any kind. Some of the larger stations could cover a handful of states, and they were doing their best to inform the public as best they could.

After a severe power surge in the 1980s, the federal government required all power companies to connect their systems together in an effort to make the power grid more flexible and resilient. If one power station went down, the surrounding stations could pick up for the reduction in power fairly easily.

In recent years, the power companies figured it would be much simpler if they just monitored each other's power production on secure lines via the internet. When the internet went down due to the satellite loss, the power companies could no longer monitor each other or even many of their own power systems. All power companies went to emergency power production but were having trouble evaluating what the demand was and what their neighboring power company was producing. They could see the demand rising on their manual systems, so they would feed more power into the network, which took time to spool up.

This is what the adjacent power companies 50 or 100 miles away would do at exactly the same time. This would spike too much power into the system, which they could all 'see' after they had powered up the generators now feeding too much power into the system. A mega power plant couldn't just turn the power up or down at a moment's notice; it took time to increase or decrease power production.

It didn't take long before they were just having to guess how much they should produce to keep the system running. After a few hours, it was like watching a badly out-of-balance tire trying to keep up with a speeding car as the tire tore itself apart. Breakers were blowing all over the national grid, putting more variability into the system, which meant more guessing by the power producers, leading to more blown breakers and more variability, and eventually, they had to all-power off to save the power plants from being completely destroyed.

In New York City, Boston, Atlanta, Miami, Philadelphia, Chicago, Detroit, Los Angeles, San Francisco, and Houston, traffic lights flickered, then went to blinking four-way red, and then off. Car accidents were common, adding to the already stressed-out police, fire departments and hospitals. Within hours, the continental United States and Canada (which had tied into the US power grid) went black. And it was August. The same fate occurred all over Europe. Traffic crawled to a standstill. Generators kicked in at fire stations, police stations, and hospitals, but most facilities only had enough generator fuel for two or three days. The president declared martial law and a 7:00 p.m. curfew over the radio. The continental United States was in chaos, as was much of the world.

Brad was able to get a call through to his father-in-law. His wife had arrived safely, thank God. The 40-mile trip had taken her six hours, and that was before the power went down. Brad heard NASA's generators kick in just as his phone line went dead. At least he knew she had gotten herself and the kids there safely. People couldn't get gasoline now that the power was off, and the few stations that had generators—mostly in the states affected by hurricanes—were overwhelmed and only took cash. Fights broke out. People turned to abandoning their cars where they ran out of gas. Trucks couldn't get through to make deliveries, and some truck drivers were mobbed in the larger cities before arriving at their destination.

Brad's phone rang, making him jump. He realized that it had been unusually quiet. As he turned on his headset, the display showed that it was the director. Brad sat up, unconsciously looking around Mission Control and at the white, oversized monitors that were supposed to be showing the ISS. "Brad Brown."

"Yes, sir. Conference room G. Right now."

15. Conference Room G

Conference room G was where joyful events and announcements usually took place. Brad was sure today would be different, but he was hopeful that at least someone might explain what the hell was going on and help them find a way out of this mess.

The conference room is like an amphitheater, set up with a large table at the front and a small lectern in front of that. There were too many uniforms to keep track of everyone. He recognized his boss and many of the faces at the large table. The man at the lectern was the president of the United States, POTUS himself, with no formal introduction, just standing there getting the meeting underway.

"Ladies and gentlemen, take a seat. I have a few things to say before General Latimore of the United States Airforce brings us up to speed on a few items. First, let me preempt what I am going to say by telling you that this is an unexpected stop. I do not have a formally prepared statement for you, but what I can say is this: What happened four days ago was an act of war. Our generals, commanders, and armed forces around the world are diligently sniffing out the individual who did this. General Lattimore is riding with me on Air Force One as we review our options and coordinate our positions with the leaders of other governments. As you know, this situation has led to more deaths than can be counted and unspeakable pain, suffering, and total disruption around the world. It is the act of a despicable coward, and we are diligently pursuing the guilty party. We will make sure they are found, and they will be brought to justice. Trust me when I tell you, they will be found. With that, I give you General Latimore."

As the President stepped away from the dais, General Latimore took his place at the podium. "Thank you, Mr. President. Ladies and gentlemen, I am going to start with some information you probably already know. Just bear with me as I work through this so we're all on the same page. Four days ago, two satellites were taken out by what appears to have been anti-satellite weapons or ASATs. We do not know who or what country is responsible for this aggressive hostility, but their action in destroying these two satellites, and in the specific manner or design of their action, appears to have been specifically aimed at creating enough debris to start a chain reaction—or domino effect—destroying even more satellites. This reaction, multiplied by the previously existing space debris, continued until the full Kessler Effect took place. It appears whoever is responsible for this situation packed some unusual material into these rockets. They were carrying a payload of something akin to steel ball bearings, or iron filings mixed with other materials like shredded aluminum foil, to intentionally cause a huge static charge. We don't know all of the different types of materials used, but the debris field was definitely intentional, and at this moment, it is surrounding the globe. The debris is moving through space at orbital speeds, bumping, rubbing, and gyrating in and around all of the now thousands of debris fields, each containing millions of pieces of debris and creating its own static charge. These static charges are strong enough to absorb any electromagnetic signal we try to pass through, either outgoing or incoming, thereby effectively cutting us off from radio communication or any kind of telemetry. As you know, that means no communication with ISS, or MMI—with its twenty-one astronauts and any ongoing space operations—or satellite communication. Each missile was packing an extensively enhanced, explosive warhead, creating a larger explosion than you would normally expect from a high-velocity impact with a satellite."

Whispers were heard as others looked around and darting eyes were met with concerned looks.

"Some of our satellites that are ... uh ... were designed to measure certain kinds of explosions alarmed right before we lost track. Again, now with all of the satellite unavailability, we are having a heck of a time gathering information or even communicating with ourselves. The domestic phone lines are backing up, and the international deep sea or buried cable doesn't have the capacity for any communications beyond the current emergency transmission load. These lines are filled to capacity. Hampering our ability to correspond with ourselves and other countries. The investigation is still ongoing. The results of which will lead us to the guilty parties involved. The last contact with the ISS was that they were engaging Pre-determined Debris Avoidance Maneuver (PDAM) to a higher orbit and hunkering down for possible debris collision. We also know the Russians were able to scramble into their Soyuz. We know they attempted reentry. We lost radar and radio communications with them shortly thereafter. Nothing further is known about them at this time, and Russia has been quiet on the subject. May God be with them. We do not know how long this static charge will remain as a barrier to communications. We do not know the extent of damage that has occurred to the ISS or the condition of the NASA astronauts aboard. We can only assume that they have taken refuge in their Soyuz capsule and are riding it out. However, this is conjecture, as we did not receive a transfer of command. May God be with them and their families. According to Director Adams, MMI can function autonomously for an extended period of time until communications are restored. The United States Air Force and NASA are continuing to evaluate the situation, which is a fluid and ongoing dynamic. To fly a manned rescue mission to the ISS right now would be suicide. If you get a chance to go outside at night, you will see that the aurora borealis is significantly larger and more colorful than it has been in recorded history. I doubt anyone in Houston has ever seen the northern lights. Well you can now. You will also witness hundreds of 'shooting stars,' which are pieces of the debris descending through our atmosphere. This is evidence that the debris field is being drawn into Earth's atmosphere and burning up upon reentry. So, the debris field is being cleaned up slowly through natural means. Unfortunately, our initial estimate is this 'natural' cleanup will take years before we'll be able to use satellites again. This is a classified meeting, so you are not to repeat any of what I am going to tell you now. The fact is that there is no man-made way to clean this up. We continue trying to evaluate the altitude, longitude, and latitude of the debris fields and the different directions of their travel. However, at the moment, the amount of static charge being produced is defeating most of our efforts. Currently, the United States Air Force, NASA, and some different private contractors are cooperating to coordinate a rescue capsule. This is an Air Force mission due to our current state of war. This capsule is not like any capsule ever used or even considered before. We are covering the crew capsule with case-hardened steel designed for high-speed impacts. Normally, this type of material would be armor on a tank or battleship. The extra weight adds dimensions to this mission, which is challenging, as I know you are familiar. It will launch out of Cape Canaveral. It will be a fully autonomous rescue mission to the ISS, using one of the starship boosters that had already been produced as a backup for MMI. Unfortunately, this will take a couple of weeks to assemble this "hardened" capsule, even working twenty-four-hour shifts, seven days a week. We will get it done correctly and as quickly as possible. Unfortunately, with sincere honesty, I do not expect it to survive the journey. The statistical odds and dynamics of a piece of debris traveling at twenty thousand miles per hour or more, encountering our rescue vehicle, are, at the moment, very high. You all understand the types of physics, momentum, and speeds involved here, and you know the situation. We are taking a roll of the dice. But as far as we know, we have three stranded astronauts on what is left of the ISS. We will not let them down. We have commissioned different telescopes to get eyes on the ISS. They have been able to visualize it and report that it 'appears' to still be in one piece. Altered, but in one piece. There is one Soyuz capsule docked at the ISS. We trust this is where the NASA astronauts reside. The current orbit and attitude of the ISS appears to be undefined and irregular. We are definitely in the middle of this. I do not expect it to get any worse. What we can do, and must do, is carry on. Carry on doing our jobs. You, me, your staff, and my staff are required to get us out of this mess. We did not make this mess, but we are the only ones that can get us out of it. We know the situation will change. That is why I, the president, and the whole world needs you to carry on. This is a dynamic and evolving situation, and the world needs you to be at your best and be sharp and attentive. We are all depending on you right now. We all have families and loved ones we'd rather be with at this moment. Do what you can from here to be of assistance to them. It is best if family members shelter in place. I am directing a small contingency from the Air Force Reserves to set up tents and cots and provide some provisional meals for you during this national emergency. The tents will most likely not be air-conditioned. This is a place for staff to get some rest, if needed, and have a bite to eat. We do not know how long this situation is going to persist, so prepare yourselves for an extended period of time before we get this sorted out. I will take any call through my office via landline. The nation needs you. Stay sharp, carry on, and may God be with us and the crew of the ISS."

As soon as the meeting concluded, Brad approached the front table. "Excuse me, General. I'm Brad Brown, CAPCOM, MMI."

The general extended his hand. "Yes, Brad, I believe we've met before."

"Yes, we have."

"What can I do for you?"

"I was just wondering about the crew capsule rescue mission," Brad said. "We have twenty-one people on Mars unable to get to the supplies we sent up for them in advance."

"I didn't know that. Why can't *Starship X* get to their supplies?"

"Because we never expected to not be in communication with them throughout the mission. There would be the expected communication lag times or delay due to distance, but no mission in history has not been in constant, uninterrupted communication with CAPCOM or Flight Control that wasn't previously planned. This mission, as you probably know, is much more dependent than we have ever been on automated systems. As soon as *Starship X* landed, we were scheduled to send a code to each starship, which would bring them out of hibernation. It was merely a battery-saving function, placing them in hibernation. We really didn't need to implement the hibernation function at all. We were in the middle of sending the code that would initiate their systems and bring them out of hibernation, right when we lost our ability to communicate with them, effectively isolating *Starship X* from their own supplies. Who could foresee long-term communication disruption such as this, with all of the redundancies we have in place? We surround the entire world with antenna, nine-meter, and thirty-meter dishes with ten different bands and frequencies, all with their own power back-up supplies to ensure redundancies. You can't say we were unprepared. However, I am afraid that *Starship X* and her crew can survive for only about ten more days without those badly needed supplies. Without them, they may not have access to basic needs such as food and water. I don't know if there's anything we can do, but if we come up with something—anything—can I get your support?"

"Of course. Absolutely." General Latimore stared at Brad for a few uncomfortable seconds, then shook his head as he looked down. "Fuck! You have my number and my complete support. I will do anything I can."

"You've been busy and have a lot of responsibilities. Thank you." Brad shook the general's hand and walked away, neither feeling any better after the conversation.

16. The International Space Station - Earth Orbit

At 12:25 Zulu, the alarms starting ringing and didn't turn off. As Commander Amhurst turned his head mic back on, he was met with an unusually serious Andrew Boynton, acting CAPCOM at the moment.

"CAPCOM–ISS," Andrew said.

"This is ISS CAPCOM," reported Commander Amhurst.

"ISS, this is CAPCOM. This is not a drill. Repeat, this is not a drill. How do you copy?"

"Five by five, CAPCOM," Commander Amhurst replied.

"Roger, five by five," Andrew said. "There have been two satellites destroyed in orbit. We're tracking debris heading your direction. There's no time for a coordinated evolution to a higher orbit for ISS. We will be initiating PDAM in four minutes. We expect debris arriving at your location in approximately twenty minutes and counting. We'll monitor both the debris fields and your status, with updates every thirty seconds, starting at T-minus three minutes."

Astronaut Giulia DeSantis from Cosenza, Italy, happened to be in the cupola enjoying the beautiful 180-degree view out of the largest windows ever in space. She had been enjoying the fantastic scenery of Earth and trying to spot her home country as the alarm sounded.

She thought it ironic as she looked around. *Here I am sitting in Tranquility node, with an alarm destroying any sense of tranquility I may have had.*

As she quickly rolled the shutters down over the windows, Commander Amhurst came on over the PA. "This is not a drill. Repeat, this is not a drill! We've gotta go people!"

According to the Policies and Procedures Manual, securing each of the different compartments and closing the hatches takes about seven minutes. Cutting some corners could reduce it to four minutes, and that is exactly what Giulia was doing at this time. Speed was of the essence, and she was going as fast as she could. All of the shutters had to be closed first to protect the windows of the cupola.

Commander Amhurst turned the radio selector to "all stations, speakers on," so everyone could hear CAPCOM progress with the updates and countdown.

At three minutes after the alarms began ringing, CAPCOM came back over the PA. "ISS, you are a go for higher orbit boost. Immediately go for PDAM."

Everyone heard Commander Amhurst repeat, "We are "go" for PDAM. Roger."

CAPCOM–Houston came back. "Initiating now."

Giulia closed the final hatch on her end of the ISS and was about to head to the Russian Soyuz capsule assigned to them by NASA. As she saw Henry Wilson, the other American astronaut aboard, buttoning up his side of the station, she thought the Russian cosmonauts should've been heading to their Soyuz as well by now.

Suddenly, her face hit the newly created "floor" rather abruptly, and all of the unsecured, miscellaneous tools also found the floor making a clanging metal sound as she protected her head from being slammed into the lip of a bulkhead. The station was groaning and making popping noises with the sudden rapid acceleration. It was as if the station were complaining in great metallic groans and whines. Giulia didn't like it at all. It was impossible to stay on her feet with the station moving, so she crawled to the Soyuz capsule banging her elbows, knees and shins along the way.

Commander Amhurst continued monitoring the station's vitals. He had secured himself to the floor with a Velcro belt over his socks allowing him some stability.

"Come on!" she yelled at him as she crawled over a laptop bouncing around the recently created floor.

CAPCOM came back over all stations, PA. "CAPCOM–Houston debris field tracking reports possible impact tracking towards your location. Impact probability, ninety percent. ETA of debris field within thirteen minutes."

We are never going to get our space suits on in time, Giulia thought as she stripped off her uniform and put on the diapers and undergarments required under a space suit. *And I just finished my second cup of coffee.* That thought suddenly seemed unimportant and rather strange, considering all of their lives were now in jeopardy.

Because of the time constraint, she left off the inner gloves and pulled out Commander Amhurst's space suit trousers, then set them by the hatch before donning her own trousers.

"There's not enough time!" she said.

"CAPCOM–Houston," the PA sounded loudly. "Ten minutes to impact threshold."

Commander Amhurst appeared at the door, pulling on the bottoms of his space suit, only too aware of the time constraints. He was cutting every corner he could, and even then, he knew he was out of time. "Where's Henry?"

"He's already inside the Soyuz getting it ready."

It usually takes forty-five minutes to put on a space suit, and that's in a zero-g scenario with one other person helping. The space suit weighs about 300 pounds on Earth, but virtually nothing on the orbiting ISS where they normally experience microgravity or zero g's. They were now experiencing about one half-g now due to the PDAM thrusters, which also meant the space suit now weighed about 150 pounds.

Commander Amhurst heard the thrusters still running, and he could tell the ISS was still accelerating, which meant this was an emergency of unprecedented proportions. NASA was pushing the ISS as far and fast as she could go. In other words, they were in deep shit.

He wondered how Henry already got his suit on so quickly. The thrusters on the Space Station are meant for small orbital changes or attitude adjustments, and Commander Amhurst was pretty sure they had been running nonstop for seven or eight minutes now. He wondered how much longer they could run before completely running out of fuel. As if on command, the thrusters kicked off, and zero g's once again returned to the Space Station as it moaned and complained like never before.

CAPCOM came back over the PA once again. "Five minutes to impact threshold."

Commander Amhurst tried to wriggle into the top half of his space suit. "Thank God the gravity has let up."

He saw Giulia struggling with the bottom half of her suit and stopped to help her. After listening to her complaints, they heard her suit snap together, and he returned to his own.

"Three minutes to impact threshold. You should be in your space suits at this time." Capcom informed them.

As Commander Amhurst struggled with the top half of his spacesuit, Giulia attempted to help him.

"Don't," he pleaded. "Just get in the Soyuz."

"What about Henry?"

Confused, Commander Amhurst looked inside the Soyuz. Henry was busy getting it fired up and not wearing his space suit all. He hadn't even started putting it on.

"Two minutes, thirty seconds," announced CAPCOM.

Commander Amhurst froze and stared at Henry, realizing nothing could be done now. Henry had made his decision. *Let's hope we can all survive this.*

"Two minutes," announced CAPCOM.

Giulia finished with her helmet and gloves and needed to get in her seat next to the window. The commander's seat was in the middle, but he was still putting his helmet on. She began helping him when they heard what sounded like hundreds of pieces of buckshots strafing across the station.

"Get in!" Commander Amhurst shouted. "Get in!"

Giulia shook her head. "You're the commander! Take your seat!" Giulia felt she wasn't really being brave as much as practical. *The commanders seat is in the middle.*

Commander Amhurst grabbed his helmet and gloves and climbed into his personally designed and molded seat next to the unsuited Henry.

A low, hissing noise grew in intensity as CAPCOM came back over all stations, PA. "Sixty seconds."

Commander Amhurst shouted, "The debris must have arrived ahead of schedule. Shut the hatch!"

Giulia did as commanded. As soon as the hatch door was sealed, they heard another round of strafing, along with loud bumps and crashing noises from something skipping off their Soyuz capsule and taking the radio antenna with it. A loud hissing came from somewhere inside the ISS, then it dissipated.

From the loud speakers inside the Station, they heard CAPCOM continue with the countdown and debris track during and after the onslaught. Then they heard CAPCOM's repeated attempts at hailing them.

Commander Amhurst attempted to hail CAPCOM, with no response. "We can hear them, but they can't hear us.

Henry agreed adding, "Yeah, our radio is not working We are hearing the ISS radio. That last round must've taken our antenna."

More debris pelted the station and sounded like handfuls of rocks being thrown at a passing car on a highway. The Policy and Procedure for Emergency Evacuation of the ISS states that when an astronaut makes it into the Soyuz with their space suit on, they should lock the hatch and change their call sign. Disembark from the ISS when they feel it's safe, or when they feel they have no other choice.

The Russians decided their best option was to disengage after the second round of strafing began. The NASA astronauts violated the policy for emergency evacuation on a few different fronts: they didn't get into their space suits, they didn't change their call sign, and they didn't detach from the ISS. As a result, no one knew exactly where they were or what condition they were in. Many were too afraid to guess.

17. Mars - Sol 5

As soon as Carl was back from his evaluation of *Starship IV* Commander Carrigan grabbed Carl's attention and motioned him aside. "Can you spare a few minutes of your time?"

"Of course."

They made their way to Commander Carrigan's study. Carl had an idea as to what the conversation was going to be about.

"Thanks for meeting with me, Carl."

"Of course. You want to know how we're going to get the supplies out of the remaining starships, right?"

Commander Carrigan evaluated Carl for a moment, fully appreciating the professional in front of him. "Why, yes, Carl. That, and some other items I'd like to cover with you. I guess you understand our dilemma well enough. Please tell me you've been considering the situation and may have some suggestions."

"I have. And I think I see the dilemma fairly simply. The starships haven't been activated, therefore, the access hatches remain locked, and the cranes won't function. We do not have any kind of remote control to activate them, and we have no way of getting to the access hatches a hundred fifty feet up. And now, neither CAPCOM nor MMI can activate them because we've lost communication with them, and so have the starships."

Commander Carrigan smiled and shook his head at the simplicity of this strange explanation and Carl's short and matter of fact response. They both knew that any supply issues this far from Earth could mean anything from slight discomfort to unspeakable pain and suffering—and even possibly a death sentence for the entire crew. "In short, yes. As our chief engineering specialist, please tell me you have an answer or two for our situation."

"According to my inventory records, we only have two twenty-four-foot folding ladders, and a hundred-and-fifty-foot drop from one of those hatches to the surface would be deadly."

"I agree. First things first though. I ran an inventory and status report this morning which showed our oxygen consumption is above the expected usage rate. Our oxygen tanks are getting somewhat low. Do you know why we may be over-consuming our oxygen supplies?"

"With most items and services we usually have a twenty percent margin which would be thirty-six days of safety margin built in at normal usage rates. However, we need to set up the MOXIE."

"It's on the schedule for tomorrow's activities." the commander injected, rubbing his face with his hands as the stress was becoming evident.

"Right." Carl answered, "Well it appears that the multiple EVA's that we have been forced to conduct due to the fallen *Starship IV* have been consuming more oxygen than expected. At least that is my original evaluation. Not only have we been conducting more EVA's, but they have been longer and harder and more stressful than expected."

At this point the commander interrupted, "Excuse me Carl. You mean more stressful than being 35 million miles from home, separated from friends and family for years. On a planet that has no breathable atmosphere. In space suits that, if torn or damaged, could immediately render death? Where we are being pelted with Cosmic galactic radiation. More stressful than that?"

Carl quickly glanced up at the commander and realized he was joking. It was time to loosen the mood he guessed. With a smile he continued. "Well, in a word, Yes". They both had a quick laugh. "What that means is that we're out consuming the scheduled usage rate of our breathable air.

Today I realized that our projections don't include all of these heavily crewed, multiple daily EVAs. Utilizing the equalization room two or three times a day. As you know commander each EVA pumps five hundred cubic feet of breathable air out of *Staten* into the atmosphere of Mars. That is a significant amount of air. Now, you have a tired, stinky, sweaty crew breathing off more CO_2, using more shower water, and we have to reverse the scrubber bottles they were breathing from while on the surface while exerting themselves. So, yes, that twenty days of margin NASA gave us is being consumed at a much faster rate than originally projected."

"I monitor the O_2 percentages every day, and they are always around twenty-one percent."

"I monitor it constantly. That's why we are always near perfect on atmospheric conditions on *Staten*. Carl responded"

"I understand that and am appreciative of your attention to detail, Carl. Given the circumstances, your efforts and dedication help me to feel more comfortable given our unexpected situation.

"Thank you, commander."

"You're a good man Carl, I honestly wouldn't know what to do without you." the commander added.

I've been preparing to set up the MOXIE (Mars Oxygen Electrolysis Unit). I actually just finished my assessment of the numbers late last night, but I didn't understand it until I was returning from the EVA just now. I could see that my numbers weren't matching the consumption rates that I was seeing over the last two or three days. That is what the report you ran informed you about. I couldn't figure it out until I was in the equalization room coming back from this EVA."

As the Commander drew silent, Carl knew not to interrupt his train of thought. It was as if he were engaging large gears as he was juggling priorities and considering possibilities.

"So, the extra EVA's is why my report is showing our oxygen tanks are being consumed too rapidly. Okay, Carl. Tomorrow, there is no higher priority than getting that MOXIE unit set up. I will get you all the help you need," the Commander said as he looked at the roster for tomorrow. "What do you need to get that unit up and running as fast as possible? We cannot take any chances with our oxygen level."

"Myself and three other people," Carl said. "And we should have it set up and working in about eight hours, maybe less."

Commander Carrigan thought for a few seconds, a concerned expression on his face. "That is two long EVAs. If you think you need more people, I will give you more personnel. If you need four or five let's get it done tomorrow. Many hands make lite work. How many people do you want?"

Carl nodded. "I think four is fine. I'll start setting up the gear for tomorrow right now."

Thinking to himself again Commander Carrigan paused, then asked, "How long will it take the MOXIE unit to return us to twenty days of residual O_2 reserves or more?"

"The specifications show that the unit will produce enough pure O_2 to start refilling our pressure tanks at a rate of approximately two times our daily usage. Utilizing the specifications for the MOXIE, we should be back to a twenty-day safety margin in ten days . . . depending."

"Depending on what?"

"Usage rate."

"Yes, yes of course. On another subject, do you know if we have any rope? I mean, I know we do. Do you know where it is and how much we have, just off the top of your head? We may be needing all of it in the near future. I haven't had time to do an inventory search."

"We have some here on *Staten*. I'll see if I can locate it. Actually, I think I know where it is. For the rest of the rope, I'll have to look it up on the computer. What do you have in mind?"

"Strong enough to lift a person safely, I guess. Two hundred fifty kilograms or so. Life safety line."

"Length?" Carl asked.

"I'm not sure. Maybe five hundred feet. Can you locate how much we have on *Staten*, *Starship IV*, and starships I and II? I am not sure what I'm thinking of at the moment. What are your thoughts around getting the hatches open on the other starships?"

"I've been thinking about that."

"The nylon netting used to secure the inventories and supplies are all secured to the aluminum framing attached to the inside structure of the starships, correct?"

"Yeah," Carl said.

Commander Carrigan had an idea. "If we disconnect all of the nylon cargo netting and remove all of the inventory in Starship IV, can we remove the aluminum framing and weld it together in the shape of a hundred-fifty-foot ladder or scaffold and maybe use the cargo netting to help reinforce the scaffold or secure it to the starship or act as a safety net or guard? You know ... wrap it around the scaffold or something?"

"Hmm ... you know, I think we could. Let me think about that and come up with some ideas. It would take a few days, but I think we can. I have electric grinders with cutting blades and aluminum welding torches and some other equipment on Staten that we will need. We could use the Mars car to haul it over to *Starship IV*, and I could do the work on the surface outside. I would just need power."

"Yes, well we haven't off-loaded the Kilopower nuclear generator yet. We have just been too busy. We will need it for the MOXIE tomorrow anyway. Please draw up any ideas you may have for the ladder or scaffold and we'll prioritize that as soon as Starship IV is emptied, and we take a day off."

"Take a day off?" Carl's face clouded over not understanding.

"Yes, after the MOXIE is up and running along with the Kilopower, we're going to take a day off. We will have been working hard for nine days straight by then. It's time for a crew day off. Off-loading all of the equipment on Starship IV and then assembling the scaffolding is going to be a lot of hard work on the Martian surface. And more heavily crewed EVA's. And after that water lines are going to be a major priority, I only see three thousand feet on *Staten*'s inventories and we are getting low on water as well.

"Yeah," Carl said, maybe a little concerned. We are low on oxygen and water. he thought, wow!

"Okay, I'm going to assemble the crew for an update. Yesterday, the drinking water tank was noticeably low, but O_2 is our priority, and there's nothing we can do about water until we get to some of our supplies and the water lines. It looks to me that we can go for a little while. I will have to start putting restrictions on water usage in a few days. No more showers after day nine, until we get to those water lines. Do you agree with my assessment there?"

Carl thought about it for a few moments. "Well, the fuel cells are working at a reduced rate, but they are still working. The drinking water levels are definitely getting low. We will have to deal with them eventually as expected, which includes service on the hydrogen fuel cells, but I won't be able to get to that for a while. Unless I do it on the day you have us scheduled off."

"No thank you Carl. You need a day off too. I know servicing the hydrogen fuel cells is a big job. And you are going to be very busy on these next multiple EVA's Let's set up the MOXIE and the scaffold as we discussed. I need to call Team Staten together for a quick update. I'm sure they're wondering what we're going to do."

As the announcement went over Staten's all stations, PA, Commander Carrigan felt a little relaxed for the first time in a while. I don't know why I'm feeling relaxed he thought to himself, we are still in a rather precarious position. Prior to the meeting, he made his way to the helm and pulled Steven to the side to bring him up on the next EVAs and the plan to use the aluminum framework inside *Starship IV*.

Steven listened intently. "Commander, I have just been in contact with the ISS.

18. Soyuz Capsule

Commander Amhurst, Giulia Di Santis, and Henry Wilson were on their fifth day of sitting inside the Soyuz capsule. They had spent the bulk of their time sleeping and talking, but the situation was intolerable and possibly life-threatening. Sitting anywhere for an extended time could lead to deep vein thrombosis also know at DVT's, meaning blood clotting from lack of movement. It could lead to stroke, heart attack, or a blood clot in their lungs.

No matter how much they discussed their safety in the in the Soyuz versus the ISS, it didn't matter they could no longer stay. They went over every possible option they might face if they went back into the ISS. It didn't matter. Nothing changed the fact that they had to get out of the Soyuz. The situation was critical. If they stayed much longer, they would die. Simple as that.

Henry had been able to get his space suit on with much difficulty and inconvenience for everyone after they had closed the Soyuz's door. It's not like they were in any rush. He was able to get it on eventually with Commander Amhurst's help, and the zero g's they were once again experiencing.

There were a few supplies in the Soyuz since it usually reentered the atmosphere and touched down within about four hours after undocking. Six bottles of water and six biscuits (the emergency supplies in case the Soyuz was blown off course) weren't going to last them for long, and they all knew it was only a matter of time before a decision would be forced upon them.

The only information they had was that the banging and smashing they'd initially heard had diminished dramatically. At the last minute, Commander Amhurst had slowly turned the Space Station long-end toward the Earth, which put the Soyuz capsules facing directly away from the Earth. He had hoped that most of the debris would be heading their direction from low earth orbit. Turning the Space Station would hopefully put the ISS between the oncoming debris and the Soyuz, protecting the Soyuz from the bulk of the debris—somewhat like the blade of a snowplow.

He knew there were a lot of presumptions in his logic, but they were out of time and information. It's also why he didn't immediately detach and begin a return to Earth. If there was a debris strike, he wanted as much protection between it and them.

Commander Amhurst grudgingly repeated the energy formula in his head, and he couldn't stop it. The amount of energy delivered by an impact exponentially increases as the speed increases. For anything to stay in orbit, it has to have a minimum speed of 17,000 miles per hour or it will re-enter the Earth's atmosphere. 17,000 miles per hour is fourteen times faster than a bullet exiting the barrel of a gun, which is about 1,200 miles per hour. A direct hit by that bullet at orbital velocity would go straight through the Space Station before anyone even knew that it had been hit.

Another key factor in the equation was how much the projectile weighed and whether or not it was a direct hit. A glancing blow may skip off, and by Commander Amhurst pointing the Space Station towards the Earth, he was able to lower the profile of the Space Station, thereby increasing the chances of a glancing blow. In recent years, NASA had the astronauts install Kevlar padding on the outside of the ISS to help reduce penetration from flying debris or micrometeoroids.

It had been five days, and they had been effectively out of water for over twenty-four hours. After they emptied their first bottles, they rationed the remaining bottles. Giulia was the only one wearing the special diapers designed exactly for this kind of situation. The batteries were running low on the Soyuz, as well. When they considered that the batteries could be too low to launch the parachutes, upon re-entry they agreed they would open the ISS hatch and hope they could formulate some sort of survival on what was left of the station. They needed to act before the dehydration started sapping their bodies of energy or they began having physical problems from sitting for so long. If worse came to worse, maybe they could dive back into the Soyuz and risk the tired batteries for a return to earth.

After much debate they decided Giulia would open the hatch, and when she did, there was only a slight rush of air into the ISS. Giulia turned her space suit lights on and saw stuff floating around and alarm lights on the navigational panel. Either the evacuation alarm had silenced itself, or the battery running it had died.

She crawled in and floated to the internal atmospheric panel, where it read 16% O_2. Not perfect, but good enough for now. The atmospheric pressure gauge was at 9.6 psi. Not perfect, but survivable. Maybe they could find the hole or holes and patch them.

As she slowly took her helmet off, she purposefully exhaled and waited for her ears to stop popping and her sinuses to equalize from the pressure change, she turned around and shouted, "Welcome aboard! We are now at 12,000 feet above sea level!" The corners of her eyes were bubbling making it look like she was crying.

Anything was better than being stuck in that damn Soyuz another minute!

Giulia shook her head to help her sinuses equalize then helped the guys get out of the Soyuz and smiled at her partners. "I love you, guys." And she did start to cry with tears of joy.

The crew shared a few group hugs, thrilled that they were still alive and in one piece.

They backed each other up as they slowly started taking inventory, making sure not to get too far away from each other since they didn't know what the next big issue or rude surprise might be. There had been no debris strikes for a while now. Giulia tried to remember how long it had been.

The strafing seemed to come in groups or bunches at a time. They concluded that it was at least four hours since they had heard the last impact, and the last two didn't seem as bad as the previous five days had dealt them. When the ISS was at its typically prescribed distance of 250 miles from the Earth, they would orbit around the Earth every ninety

125

minutes. But now they had no idea what their orbit was or even how far from the Earth they were.

PDAM, had been activated, which had a predetermined orbit of 500 miles, but with all of the excitement, the commander wasn't able to verify anything relative to their orbital positioning before entering the Soyuz capsule. PDAM had never been actuated before. They decided they had better start looking for leaks and try to get the O_2 levels and atmospheric pressure back to more reasonable levels. They silently floated around the Unity node, listening for the quiet, telltale hissing sound that would give away the location of any leaks.

It was decided they wouldn't open hatches to any other nodes until they had secured the leaks in the Unity node, then they would discuss venturing farther at that time. As they came across the occasional tiny leak, they would patch it up with a special two-part epoxy, and then cover that with the universally used heavily reinforced duct tape.

They had three normal air bottles and two pure O_2 bottles in storage. They cracked the valve of one of the air bottles and opened it slightly. After fifteen minutes of listening to the precious air mixture escape from the tank, the pressure started to rise inside the Unity node. They left the tank valve open until they were approximating the altitude of Denver or around 5,000 feet in altitude in their artificial atmosphere in the vacuum of space.

Commander Amhurst checked on the batteries, which appeared to be undamaged. The breakers had blown, or the fuses had tripped in the fuse box. They didn't want to turn it on and start a fire somewhere, so they worked together to jury-rig some lights and then reconvene.

They had been in their space suits for five days and desperately needed to remove them and clean themselves up. Giulia was the only one who had put the diaper on, and she felt a little guilty that she had no significant personal cleanup responsibilities other than to discard it. They hadn't eaten or had much to drink, which somewhat limited the cleanup job for the guys.

Giulia checked the water stores and removed some drinking pouches, which they all enjoyed.

The next big decision was in front of them: What do they do next? The ISS was filled with cameras so Ground Control could monitor the different experiments and activities. The doctors and physiologists all wanted to "see" them working out or doing various duties and experiments. Many of the ISS astronauts in the past had felt like guinea pigs, constantly under the microscope, with heart rate monitors and respiratory monitors giving the doctors feedback on how their patient was doing constantly day or night, twenty-four hours a day. It was so invasive that an entire crew turned all of the cameras off, disconnected all of their bio-monitors, and took them off. NASA decided that the doctors had gone a little too far, and the monitoring was toned down significantly after that. The cameras on the ISS were off at the moment. The radio had no power, either.

While the space suits were drying by the hatch, after being cleaned as best they could, they ran a new power line to the radio, and with some ingenuity, they were able to get it to fire up. They tried to hail CAPCOM but weren't surprised when their hails went unaddressed. They continued trying to hail CAPCOM for a good twenty minutes. The antenna was probably damaged or missing from the top of the Space Station, just as it seemed to be on the Soyuz.

As they were about to turn the radio off to preserve the house batteries, they heard, "ISS, this is *Starship X*. How do you copy? Over."

Commander Amhurst stared at the radio like he had heard his long-dead grandfather talking to him. He slowly picked up the mic. "This is the ISS. Copy, four by four, *Starship X*. How do you copy? Over."

Commander Amhurst, Henry, and Giulia looked around at each other, not believing what they had just heard. Commander Amhurst checked his watch and scrambled for a pencil to write down the time on the wall next to the radio. They looked at the radio, then at each other again, and they all started to laugh at the irony and pure joy of making contact with another living human being, even if that person was on Mars.

"What are the odds!" Commander Amhurst said. "That was *Starship X* on Mars. They probably want to know why we keep hailing CAPCOM." Commander Amhurst had an inquisitive look. "What does that mean? Maybe they heard about the satellite strike? But why are they hailing us?"

"We haven't heard from CAPCOM in five days, but I thought that's because our antenna was missing on the Soyuz. It's obviously not missing on the station, or we wouldn't have heard *Starship X*."

"So, that means that CAPCOM. . . ." Commander Amhurst picked up the mic. "CAPCOM. CAPCOM, this is ISS. How do you copy?" The seconds ticked by, and

Commander Amhurst tried again. "CAPCOM, CAPCOM, how do you copy?"

Nothing. No familiar squawk of the mic activating. No friendly voice responding.

Finally, he heard, "ISS this is *Starship X*. We copy you two by two. Good to hear someone's voice out there! Are you in contact with CAPCOM? Over."

Commander Amhurst documented the time. It had been almost twelve minutes. How could *Starship X* hear them? They must be, was forced to think about the solar system for a minute, 35 million miles away?

Commander Amhurst picked up the mic again. "Negative, *Starship X*, we are not, repeat, not in communication with CAPCOM. Due to a satellite strike, we had to execute emergency PDAM. Numerous strikes have knocked out much of our equipment. We were just getting the radio back online now. We are surprised and relieved to hear your voice. We've not been in communication with CAPCOM for five days. Over."

Commander Amhurst made a note of the time and wrote it down on the wall below the last time stamp. They decided to stay close to the radio and grab something to eat while they waited for the next return radio transmission from *Starship X*. As they sat around sucking down water again—after eating cold chicken soup, which was supposed to be hydrated and heated by the no longer functioning hot water machine. They started making prioritized to-do lists. The number one item was dealing with the offline atmospheric controls before the temperature started getting out of control. The temperature was currently a chilly but tolerable 48 degrees inside the Unity node, and they fortunately had access to more clothing. Giulia took a washcloth bath then put the undergarments required for the space suit, pants and a shirt, back on along with one of her NASA sweaters *Thank God these underwear are antibiotic.* she thought.

As they discussed their next move, the radio squawked again. "ISS, this is *Starship X*. We have not been in contact with CAPCOM for five days, as well. We are operating autonomously. Do you have any information as to the loss of comms with CAPCOM, and are you safe? Over."

Commander Amhurst looked at the time stamp he had written on the wall, then checked his watch and made another notation. He was pretty sure he was talking to Steven Henderson, whom he had trained with before Steven was selected for MMI. It was evident Steven was sitting by the radio, waiting for their transmission return every twelve minutes, just as the ISS crew were. This told Commander Amhurst that they were just as confused as the ISS crew was about the break in comms with CAPCOM and anxious to get any information they could.

Commander Amhurst picked up the mic again. "Negative on info relating to CAPCOM. We were sheltering in the Soyuz for the past five days due to the satellite strike and subsequent debris field. We know bupkis other than that. We are currently assessing our situation on ISS. Unity node is our intact domicile. However, power and atmospheric controls are limited. We are still evaluating our situation. Over."

Commander Amhurst purposefully used the term "bupkis" as a code for Steven, which was one of his favorite ways of saying, "We don't know anything." If it was Steven, he would understand it immediately.

To continue assessing the condition of the whole station meant putting their space suits back on before opening the hatch to the next node. Each node had been sealed with its airtight hatch and opening a hatch to another mode meant risking the air, the atmosphere and the roominess they were now very much enjoying.

The radio squawked again. "Hey, Commander Amhurst. You're right. This is Steven. We're in better shape than you at the moment. We all wish you the best of luck, and we will get back to you as soon as we know more. Let's set up a schedule for telemetry every day at twelve noon Zulu. Over."

"Roger that, Steven," Commander Amhurst said. "Twelve noon Zulu. We will do the same. Over."

They discussed the situation and the risks involved with opening the hatch to any of the other nodes. It was possible that some of the other nodes were wide open to space with zero atmosphere and 200 or 300 degrees below zero Fahrenheit. They could lose all of the atmosphere they were currently enjoying after the five-day nightmare in the Soyuz.

On occasion, they would watch the Earth pass by one of the hatch windows, and it looked tiny compared to its normal size.

As the Earth zoomed by, the fleeting views they caught told them little other than they were still in orbit. The other hatch window would occasionally let the sun come gleaming through at odd, ever-changing angles. But it did seem to be warming the node at times. The hatch windows into the Quest Joint Airlock or the Tranquility node usually stayed black, with occasional darts of streaking sunshine quickly passing through, but no useful visibility. On occasion, they would hear a random scrape or popping of the hull as it would roll into or out of the sunshine and something occasionally was scraping around outside. But there was little indication of what lay beyond the hatch doors or the condition of the rest of the Space Station.

19. Joint Chiefs

General Eugene Taylor looked around the beautifully polished mahogany table at the meeting of the Joint Chiefs. It was beyond comprehension that they had little more information than they had four days prior.

They reviewed the satellite feed that had been downloaded at the time of the rocket launches. The view was from four different geostationary satellites over the Atlantic and Pacific oceans. They could see from the images a deep blue sea suddenly erupting a brilliant white foam, which was the launch location of each of the rockets. Unfortunately, the focus of three of the satellite cameras wasn't anywhere near the launch point of each of the rockets, and only one satellite had the target somewhat in focus. But even it was of little help. All it showed was a calm, beautiful sea exploding in white, and some cigar-shaped cylinder belching yellow with black and white smoke and steam following it. The camera only had four seconds of the digital image before the target flew off camera. No amount of digital enhancement produced any further details, and nothing specific could be identified. It could've been made by anyone—no markings of any kind. The shape was unremarkable and generic in nature. The only reference point as to the size of the rocket was the state of the seas reflecting the wind driven wave pattern. And at the time of launch, the hydrology and ocean specialists reported the timing from initial blast to exiting of the water suggested the rockets may be approximately 60 feet in length.

Secretary of Defense Robert Thornton and the president were going crazy because no one had any leads as to who the guilty party was. There was no sign of a launch platform or vessel or extended wake or prop trail following any kind of launch craft other than a small fishing boat from Fort Lauderdale with two retired firefighters aboard.

The firefighters' stories checked out, and just prior to the launch, one of them reported seeing a "black thing" bobbing in the water around a mile off the port bow. The Coast Guard and Navy conducted a 100-mile radius physical search. The Bahamian Government supported a search of Grand Bahama Island, Bimini, all of The Berry Islands, Andros, and the busy island of New Providence, with no further findings of interest. In the Pacific, it was a different story. The location was so remote it took the Navy cruisers docked in San Diego and Pearl Harbor two days to get in the vicinity. The Airforce was on scene first and performed as many searches of the un-ending sea as possible, but they found nothing of substance. A few sailboats, six cargo vessels, a few oil tankers, along with two fishing boats from Chile were located and boarded, but no helpful results could be substantiated, and they were all hampered by the lack of usual satellite communications.

It was as though everyone forgot how to do anything without the satellites. The satellites watching Cuba 24/7 had been downloading pictures and video as usual, which resulted in nothing. no unusual activity for the previous twenty-four hours to the launch, either. Then the satellites were blown up, and all satellite signals were lost and probably would never yield anything more since they were programmed to download all of their images within twenty-four hours. The technicians said they would be overwriting their memories by now if they had survived the debris storm, which they probably hadn't, from the sound of it.

Never had it been considered that the satellites would lose contact with the ground-receiving stations for more than a minor blip, or the short time it would take to come over the horizon to communicate with another dish or antenna array. In our modern, technology-dependent world, no one had stopped to ask the question, "What should we do if we lose the satellites?" The Joint Chiefs were asking that exact question now, and no one had any answers.

The United States was blind, as were all countries overly dependent on satellite resources. The United States was in a de facto state of war, but with whom, they didn't know. All nations around the globe were now dependent on WWII technology to protect their boundaries. The whole world was back on an even playing ground, which drove each of the Joint Chiefs from the Navy and the Marines crazy. No one was getting any sleep. They were all coming up with different ideas and were busy chasing down false leads to no avail.

Each of the military branches gave a brief report on what they were doing in the aftermath. The Navy had sent out every aspect of its multiple fleets to monitor the seas and skies with penetrating radar. The powerful sixth fleet had been recalled from the Mediterranean, much to the chagrin of the Saudis and Israelis. This attack sent all military branches into a tizzy, the likes of which would probably never be seen again.

All aspects of the armed forces were on high alert, but everything was silent, as if they had been blindfolded and sucker-punched, and now they were deaf and blind. The Joint

Chiefs from each division of the armed forces speculated as to what country or player was responsible for the attack, but no one could come up with any documentation.

In an unprecedented action since WWII, each military branch was coordinating to back up each other's positions. The National Guard was overwhelmed with demand, coordinating and helping food and supply shipments get through, stopping the looting, and monitoring the curfew at night. FEMA was working with them, trying to keep the peace at home and distribute food and water in the inner cities. Chaos reigned supreme. Shots could be heard in many major cities at night. However, the good news was the power companies were reporting that if they completed disconnecting the interlinked power grid, they could start producing power as soon as they ran systems checks.

The Air Force was working with the Air Force Reserves, which rallied to the call, having been denigrated in the past to no more than a bunch of people licensed to fly together. Now they were running border observation missions, reporting anything that could be of interest to the Air Force, and the Air Force was actually listening to them. Rocket scientists brought in from NASA and every leading independent rocket corporation in the United States responded with their knowledge and expertise when questioned about the type of rocket that may have been used.

How could anyone launch in what appeared to be the middle of the Pacific and 20 miles off of the coast of Fort Lauderdale with no launch pad, platform, or barge? Could it have been a submarine launch? The Navy responded to this saying they had every submarine, domestic and foreign, accounted for, and there weren't any in that area of the Atlantic or the Florida Straits. When questioned as to how they knew that, they just smiled smugly and said, "Trust us. We know." This didn't help tempers any.

What they learned from the rocket scientists shocked them all. It turned out the ocean offered a stable and anonymous platform for a rocket launch. It was much easier than any of them thought even possible. Almost all rockets are densest at the base, with watertight and airtight fuel tanks, combustion chambers, and plenty of trapped air pockets leading toward the nose cone. If the nose cone were lighter than the base, as most rockets are, it would float upright, and if it were painted black, it would look to anyone passing by that a channel marker or a piling had broken loose in a previous hurricane.

According to the Navy and Coast Guard, there were always things like that floating around the ocean. Anyone could study the ocean currents, drop off their airtight/waterproof rocket, and get out of dodge. Twenty-four to thirty-six hours later, the drop off vessel could be 500 miles or more away, and they could launch with a high degree of certainty of hitting the target as long as the programming for navigation and the timing was done correctly.

In this day and age, the technology, save for the programming and coordination timing was not all that difficult. Many countries had the technology. Ironically, the launch code could've been sent from a satellite phone—and it probably was. So, the initial search area of 100 square miles was five times smaller than it should've been, and even that may have been too small.

It was becoming apparent that they would find no evidence trail leading to the guilty party. After days of interviews and exasperating arguments, David Burdick, the secretary of defense, suggested the Joint Chiefs were never going to find the evidence they needed. He suggested the Joint Chiefs start considering the facts from the perspective of who would benefit from such an attack—someone, or some country, that could benefit from the results of the attack, which effectively leveled the playing field.

What country would benefit from every nation being reduced to World War II technology all over again? This angle brought up some interesting conversation. Russia, or the Soviet Union, and all of the developed world was immediately removed from suspicion since they had also become dependent on satellites. As usual with the government, a joint task force was convened to evaluate the situation.

From the shifting of eyes and the telling of a few faces, a new logic and quiet awareness was creeping into perspective. A few generals started a flurry of aids entering and leaving with hurried memos. They had sneaking suspicions, and they weren't feeling good about them.

20. Mars - Sol 6

Commander Carrigan had been in his study, thinking and planning. It was 11:00 Zulu. Soon they were going to have to adopt their own clock for Mars time since a day on Mars is forty minutes longer than a day on Earth. For telemetry purposes, all communication equipment would always use Zulu. However, soon the two clocks would be incompatible. It will be tomorrow on Earth, while it was still the day before on Mars.

The sun had risen on Mars about an hour before, and Carl and his team were already on the EVA setting up the MOXIE oxygen-generating equipment. The temperature this early in the day, even during the Martian summer, was -40°F strangely enough this is also -40 degrees C. The tiny atmosphere on Mars doesn't retain much heating of the sun from the day before, and before noon, it would be above freezing again and probably approaching 50 degrees or more.

Fortunately, the weather had been friendly to them. They would have to set up the Kilopower portable nuclear generator for power production for the MOXIE and for the welding equipment as they welded the scaffolding together. Hopefully this will give them access to the other starships where two more Kilopower units were stored. As winter approached, *Staten* would soon need the Kilopower units to keep her batteries charged, as she would be demanding more power to keep her and her crew warm.

The commander wanted to be out there helping and overseeing the set-up of the MOXIE, but he needed to think. He knew Carl could handle this important work on his own.

Steven Henderson had just arrived in Commander Carrigan's office. He wanted to discuss the comms from the night before with the ISS and their implications. Steven had given him a short breakdown of what the communication over the radio had been the previous night, and they wanted to digest it a bit.

The commander had just announced the all-crew meeting was about to start when Steven had pulled him aside and informed him of his conversation with the ISS. During the meeting, *Staten's* crew were surprised at the oxygen issues, just as he had been, but they were also relieved they had an answer to the problem that would be implemented the next day. The crew was interested in the fact that they had been in communication with the ISS and that the ISS crew was safe, for now. Of course, they wanted to know why *Staten* could "talk" to the ISS but couldn't "talk" to CAPCOM. It was the question of the week, and one to which no one had an answer.

Commander Carrigan would much rather tell the crew he didn't know the answer to a question than have an uninformed crew suspicious of rumors they had heard. After all, it would be easy for a crew member to overhear the radio communication with the ISS and then wonder why it wasn't mentioned in the crew meeting, bringing them up to speed.

The crew was also looking forward to getting a day off (their first since landing) and seemed interested in the theory of how to access the other starships. The commander hoped they would somehow be able to get them to come online manually. Otherwise, they would have a lot of heavy lifting and unloading to do, and many more EVAs had been scheduled, putting them much further behind the original schedule of discovery and science. Every EVA held potential exposure and risk to crew members and equipment.

Commander Carrigan was sitting at his desk, thinking about these things and what the future may bring when Steven, who didn't look like he had gotten much sleep, walked in and took a seat. Commander Carrigan looked at Steven for a moment and immediately wondered why he looked so stressed.

Commander Carrigan started, "Hello Steven. My takeaway from your telemetry with the ISS is they don't know why they're not in communication with CAPCOM either."

"Good morning commander. Yeah, that's my take on the situation as well," Steven said.

"There's something you're not telling me."

"They had to engage PDAM and were hit with what sounds like a lot of debris. That's why they thought they couldn't talk to CAPCOM. They thought they had lost the antenna."

"Oh, my God," Commander Carrigan said grimacing. "They're lucky to be alive."

"Yeah, they are. The conversation got me to thinking and I've been going over it all night. I have an idea I'd like to go over with you."

"I'm listening."

Steven sighed. "I'm concerned that the Kessler Effect may have been triggered."

Commander Carrigan frowned. "You think there may have been enough debris to initiate a chain reaction?"

"Yeah, that's pretty much it," Steven said. "Even though we only communicated with the ISS for a few minutes, they explained to me that they had to engage PDAM.

They evacuated to the Soyuz capsule and rode it out for five days after enduring what sounded like some serious impacts. For example, they were surprised to hear us because they thought the radio antenna was missing off of the Space Station. That's why they thought they hadn't heard from CAPCOM. Now they are camped out in the Unity node with no atmospheric controls. They haven't even tried to enter any other portions of the ISS as of yet. Fortunately, it sounds like it's holding pressure so far. Imagine five days stuck in that tiny Soyuz capsule!"

Commander Carrigan's eyebrows raised, wrinkling the skin on his forehead. He sat back in his chair looking at the ceiling and paused visualizing the scenario. "My God, it must have been torture, five days in the Soyuz? I don't think I can imagine five days stuck in the Soyuz. What kind of hell must they have had to endure? Unbelievable, what is the state of the ISS?"

"We really don't know." Steven said. "They reported that they are only utilizing the Unity node. They lost communication with CAPCOM at approximately the same time we did, after a satellite strike. Did that satellite strike cause enough debris to finally cross the threshold and create the Kessler syndrome?"

"Let's say for the sake of conversation that the Kessler syndrome threshold was finally crossed. That would prohibit any rocket or spaceship of any kind from entering or leaving Earth safely, without being smashed to pieces. What does that have to do with communicating with CAPCOM or NASA?"

Steven rubbed his forehead. "That's what has been eating at me all night. The only thing I could think is that maybe the Kessler Effect is somehow charging the atmosphere, absorbing the radio signals."

Commander Carrigan squinted as he considered his response. "So, you're saying the debris is somehow creating an electric or a static charge, absorbing any radio transmissions? Any radio signal getting in or out?"

"That's all I can imagine could be happening," Steven said. "I've been thinking about it all night. Now that we know the ISS isn't receiving radio transmissions with CAPCOM, but they can hear us, that definitely tells us the problem is with Earth. We suspected that all along, but now we might know the reason."

"Ok, this is ludicrous but Steven, if this is the case, can anyone do anything about it? I mean, I'm sure that given the right amount of time, gravity will prevail. Can NASA or any other agency do something… or anything?"

"Yes, of course. Given the right amount of time—an unknown variable—gravity will, as you said, prevail. Otherwise, unless you know something I don't, your guess is as good as mine."

"So, any launch or landing to or from Earth may be jeopardized or impossible for an unknown period of time?"

"Unless someone is specifically trained in atmospheric sciences and all of the specificities related to this issue, it's anyone's guess. But Commander.... That is a possibility. One that I have been going over all night."

Commander Carrigan had to think about what Steven just put on the table—that it could take "time" to clean up. How much time? If the Kessler Effect had, in fact, taken place, could this even be a possibility? Could this situation actually make it impossible for them and the ISS astronauts to return home? If *Staten* misses their return window next summer, they couldn't attempt to return to Earth for another two years due to the huge variants in the orbits of Earth and Mars. The distances are just too great. The seriousness of the situation began to bear down on him. Commander Carrigan began to realize this might not be just a communication issue. Could this communication breakdown and possible Kessler Effect prolong their scheduled one-year stay on Mars and turn it into a three year stay… or longer? How long could it last? Could they survive on this planet for three years or more? They did have more than one year's supply of food, but only as a redundant precaution. About a total of sixteen months of food was packed into all of the starships put together. They simply did not have a three-year supply. They did have seeds and pressurized greenhouses, but for scientific research purposes primarily. They didn't even know if they could get anything to grow.

No one knew how long it would take for something like the Kessler Effect to dissipate and make space safe for travel again. *Staten* had the most highly trained crew in the history of space travel, yet they didn't have anyone specifically trained in this type of atmospheric science. He needed to do further research on the Kessler Effect and consult with Barbara Black before discussing this with the crew—when it was the right time.

Commander Carrigan was obviously shaken by the conversation. "Steven, how much have you looked into this possible scenario?"

"You know I am by no means a specialist in atmospheric sciences, but after talking with the ISS crew yesterday, I started researching it. I was up all night trying to sort through the data and scenarios and found a good bit of data on *Staten's* computer. If the Kessler Effect has taken place, as far as I can tell, we may be here longer than we anticipated. That's why I wanted to meet with you this morning. I am still following up, but we may want to reconsider our in-situ resources and the length of our stay."

"You *have* spent time on this," the commander said. "I'll start researching it as well. If we determine that it has happened, everything changes. We move from a scientific exploration mission to a survival mission. So far, all we know is that we lost communication with CAPCOM at approximately the same time that the ISS was engaging PDAM due to a satellite strike. Then they lost communication with CAPCOM as well. But why? I'm afraid that little detail may be ominous. Can there be any other reason for the simultaneous loss of comms? Let's follow up with this disturbing conversation when we have more facts or evidence."

Raising his eye brows and looking now directly at the commander, Steven asked, barely audibly "Commander?" knowing they already knew.

"I know what you're going to tell me. You're going to tell me that we already know. The only reason that the ISS could communicate with us and not to CAPCOM would be that the Kessler Effect has taken place. We know that a satellite has been hit, and the debris caused significant damage to the ISS. It's the Kessler Effect, isn't it?"

"I'm pretty sure it has taken place, Commander Carrigan, yeah. Not positive, but I'm pretty sure," Steven said quietly.

"Yeah, I am afraid there can't be any other reason. I had no idea."

"Nobody did. That's why I've been up all night researching this Kessler theory."

"Okay, Steven. Let's wait until we're as close to one hundred percent as we can get before any of these details are discussed outside of this room. Man, I hope this isn't the case, but I'm afraid you may be right. Let me know as soon as you have anything further. By the way, have you found out anything about booting up the other starships?"

Steven shook his head. "I've been trying to find out anything I can about turning on the other starships. I just don't see anything."

"Once we gain access, we can try to fire up the computers," the commander said. "If they don't fire up, we're back in the same position we are in now—manually unloading all of the supplies. We're back in the 1800s manually humping supplies on Mars."

"I'm going to get back to monitor the EVA," Steven said. "Unless you have something else?"

"No, I was going to go out with them this morning, but we needed to have this conversation."

"Yeah, we've been here for six days, we've done like ten EVAs, and I think only one of them was actually on the first week's official schedule. Crew safety is our primary concern, and every EVA so far has been an attempt to access our required equipment or install the appropriate equipment to keep *Staten* functioning."

"Well," Commander Carrigan said, "no one expected Starship IV to fall over. And no one expected that the starships would not be autonomously up and running by now. I'd like to evaluate Starship IV's stringers that we're going to remove and use as a scaffold, as well as get a look at how Carl's coming along with the MOXIE installation. We will need a back-up crew of at least two. Why don't you come along, Steven? Have you been out on an EVA yet?"

"I have to talk to ISS at twelve noon Zulu"—Steven looked at his watch—"which is almost right now."

"Go ahead and make the radio call with the ISS. We can go as soon as you're finished. I have a few things to finish up here anyway."

"Okay," Steven said. "I'll coordinate with Mary. She's monitoring the EVA."

"Sounds like fun. I need the distraction. By the way, it's not an issue if I discuss the Kessler syndrome with the ISS, is it? They may be considering it by now anyway."

Commander Carrigan paused solemnly and looked into Stevens eyes and said, "The more they know, the better their chances of survival."

"Yeah, that's what I was thinking. They're in pretty rough shape. I wonder if the Russians survived reentry. They didn't mention them at all—just that they had spent five days in the Soyuz. I just assumed that the Russians made a run for it. Imagine spending five straight days in your space suit."

The commander shook his head. "It sounds worse than Apollo 13."

"I think it probably was. May God be with them."

"Okay, make your transmission, then let's suit up. We have an EVA to go on!"

Steven thought Commander Carrigan actually sounded happy.

21. Johnson Space Center - Day 6

Brad Brown was back in his chair, now wearing a Johnson Space Center jumpsuit with nothing on underneath. After six days, his MMI team shirt and the khakis he was wearing were a mess. He badly needed a change of clothes, though he realized that if that was the worst of his problems, he was in good shape.

Brad and his team had just finished up going over the MMI computer protocols on Starship X. These were the protocols that the crew for Starship X had available to them on their shipboard computers. There was no mention anywhere in their protocols on how to start up the other starships if MMI didn't send the command. No one saw this situation occurring. This was a big oversight, and one that would not be repeated.

Brad had been working with the computer programming engineers, design engineers, electrical engineers, and anyone else that might have had some insight into the engineering and design. If CAPCOM or Ground Control happened to regain comms for a few minutes, or even a few seconds, Brad wanted to be ready.

They were transmitting the initiation orders every hour just in case one of the transmissions was able to get through. The engineers had been working on this problem for the past five days, and they concluded that there was a manual boot-up option to start the computers on the starships. It turned out that if they simply grounded the frame the computer sits in to the ship's ground, the system would boot itself up. Simple, but Brad doubted anyone would think of that.

It seemed like a technique from the 1960s—not modern-day sensitive computer stuff. Apparently, it would throw a switch that wasn't part of the computers, but the switch would initiate the power to the computers as if they had experienced a power glitch or temporary interruption. It was explained to him that it is like a manual reboot. Just unplug it and restart it.

Brad asked the programmers if there was any way the *Starship X* crew could figure this out on their own. They explained that this action wasn't part of the regular flight training, but that it should have been. And they pretty much doubted that anyone on MMI would know this or try it.

Brad's phone rang. It was his wife letting him know they had power and that she was thrilled they were going to have air conditioning that night.

"It is so good to hear your voice." he told her. "So much has gone on over the past few days."

She knew that she probably couldn't be informed about his activities at work, so she didn't ask. She reported as much as she could on the few days since they had spoken. After Hurricane Harvey, her father had some solar panels installed on the roof with a Tesla Powerwall, which was enough to run some fans, keep the refrigerator and freezer working, and keep the TV on using an old set of rabbit ears. But it wasn't enough battery back-up to run the air conditioning too. They hadn't gotten much sleep due to the heat, but at least the Powerwall was being fully charged each hot sunny day, and that kept the fans running all night.

She said the TV news was reporting that the power stations were starting to come back on, though many people had died due to the heat alone. The retirement homes and hospitals had banned the press based on medical privacy issues, but there were many rumors of a lot of deaths especially among the elderly. Many of the retirement homes and communities had too little or no generator power. Most of the hospital power generators had run out of diesel two days earlier. The assisted living facilities had no air-conditioning and no refrigeration and neither did the hospitals and they were all operating on skeleton crews. The hospitals were taking life-threatening cases only, of which it seemed there was no shortage.

The hospitals were out of many supplies. The National Guard was operating helicopters to resupply the hospitals as much as possible. Many water plants all around the US had run out of back-up generator fuel as well, so many cities and towns weren't getting running water. In some cases, they were getting water, but they were on boil water orders from the water department, if they were able to get the message. But there was no power to boil the water anyway. It was a mess.

Fortunately, Brad had suggested she take all of the food they had in their house to her father's, which helped tremendously. Unfortunately, her father had an electric stove that drew too much power, so they ate mostly cold food and cereal until the milk was gone.

Her father had started a campfire outside, which they used to heat water for coffee and pots of soup. Fortunately, he was on well water. The bread and milk were now gone, and some of their meals had been cold beans and fruit from a can, but they were doing fine otherwise.

Her father had loaded his 12-gauge and kept his German Sheppard tied up by the front door. The heat was devastating. It was 105 degrees yesterday. The news also reported that some major metropolitan cities across the US were now starting to get power. She complained that her cell phone still wasn't working. Brad explained that it wasn't going to work for a long time because the satellites were down. She was thankful he suggested going to her dad's place. It had been the right move to get away from Houston, and the kids enjoyed being with their Grandfather.

Then Brad heard the back-up generator at the Johnson Space Center turn off. Finally, it was quiet again.

She said, "Thank you, I love you." and was gone.

Brad was very thankful for the few minutes they shared on the phone. It was very relieving to know they were safe.

Brad suddenly realized that it had already been almost an entire week since they'd lost communications with the crew on *Starship X* and the ISS. What kind of hell the ISS crew was going through, he couldn't imagine. MMI had assets and supplies designed for an extended mission if required. At least their spaceship wasn't in the firing line of objects heading towards them faster than the speed of sound. Brad caught himself wondering if they could see Mars from the ISS. Of course, they could see...then Brad remembered General Latimore mentioning the rescue capsule being built out of extra thick, specially hardened steel. It gave him an idea. Was it possible . . . ?

22. The ISS - Day 6

Day six, Giulia thought as she looked at Commander Amhurst.

"Day six," Commander Amhurst said, as though he were reading her mind. "Giulia, Henry, let's have a chat. As I'm sure you are aware, tomorrow will be one week since we had to scramble into the Soyuz. I've been thinking about our situation. CAPCOM and NASA and various agencies are scrambling to try to do something to ameliorate our situation. What that something is, I have no idea. Have you guys given it any thought?"

"Henry and I talked a little about it last night while you were sleeping," Giulia said. "We were wondering if there really is anything they can do."

Commander Amhurst said, "Yeah, I have been thinking about it, of course. The answer is, we just don't know. What we do know is that we had better be ready for anything. The good news is that the impacts from the debris seem to have subsided. We need to start considering our options. We have discussed going into the other nodes, which may destroy the wonderful atmosphere that we are enjoying, so that's not a good idea. What I want to hear are your thoughts or ideas on anything else we can or should be doing."

Giulia looked at the others. "Go fish," she said.

Commander Amhurst looked at each of them. "What?"

"We have our space suits by the Soyuz hatch, ready to put on in case of an emergency. This morning we took a full inventory of everything here in the Unity node. The atmosphere is satisfactory for the moment. There isn't much we can do but play cards and bide our time. This may be the most difficult time of our journey, sitting on our hands, and trying to decide the right time to act. The debris could be missing us by a few inches or a few hundred miles for all we know. I'm sure NASA and the other agencies are working on a solution. So, we turn the radio on at twelve Zulu, talk with *Starship X*, and see if we can involve them in our decision-making process—or at least get some input from them. The more ideas we have to work with, the better. Plus, if anything should happen to us at least someone else will know what we chose to do."

"When the time comes to make a decision, we will make it at that time. Every day that we sit here, Earth's gravity is cleaning up the debris a little more than the day before. So, if we should need to take the Soyuz home, we will have a little better chance of surviving our trip the longer we wait."

Commander Amhurst and Henry nodded in agreement.

"It looks like we're going to spend some time here," Giulia said.

"Unfortunately, the longer we're here, the better our chances are of making it home in one piece. We don't have any playing cards. Do you have any other ideas?"

"Let's play two truths and one lie," Henry said.

Giulia and Commander Amhurst both replied, "What?"

"You've never heard of two lies and one truth?" He could tell from the confused faces that they hadn't. "I used to play this with my parents all the time on road trips. I'll start off, so you can see how to play. I'll tell two lies and one truth. You each get one guess as to which one the truth is. If you're right, you get to go next. If you're both wrong, I get to go again, but I have to tell you about the truthful statement and the story behind it."

Commander Amhurst and Giulia were surprised to see the typically quiet Henry perk up with his game idea.

Again, they replied, "Okay."

Henry began. "Statement number one, I had a pet rabbit growing up in Wisconsin. Statement number two, I took a flight that never existed on a commercial airline. Statement number three, on my eighth birthday, I was given the same gift by three different people. Okay, Giulia, you go first. Tell me which one is the truth."

"If I had to guess, I would say growing up in Wisconsin with a rabbit as a pet would not be unusual, but then again, getting the same gift from three people on your birthday is pretty unusual, but not hard to believe. Taking a commercial airline flight that didn't exist sounds really intriguing. I have to guess the truth, right?"

"Yes, you have to guess which one is the true statement."

"I think that having a pet rabbit is the true statement."

Henry and Giulia both looked at Commander Amhurst.

"I have to agree with Giulia that the pet rabbit is pretty realistic sounding, but the point of the game is to confuse the guessers, so I guess I'd have to say that taking a flight that didn't exist on a real commercial airline sounds pretty outlandish, so that has to be the truth."

"Your final answers?" Henry said.

They nodded.

"I can't believe you got it that easily, Commander Amhurst. I guess I need to be more subtle next time."

"How did you take a flight that didn't exist on a commercial airline? Whatever that means. You need to explain that one."

"Well, I was dating a flight attendant who worked for one of the smaller airlines. She had gotten me a 'space available ticket,' which means that if they have a seat, you can go. We had arrived quite early for the flight because she was working it. She went on the plane with her other flight attendant co-workers. She came out of the boarding ramp all smiles and informed me that the flight was canceled since the airline didn't sell enough seats. There was another flight, and the few paying passengers had been transferred to that one. She said they still had to ferry the plane to the original destination and that the captain had said I could have a seat on board while they transported the empty plane. So, I took a flight that didn't exist."

"So, I got lucky, and I'm going to get lucky again because it's time to turn the radio on. Game delay until further notice."

"Okay, but you won commander, so you start the next game."

Commander Amhurst agreed as he switched on the radio. "We don't want to miss our friends on *Starship X*."

As the three shared some water and reconstituted apple juice, they heard the radio crackle to life. "ISS, ISS, this is *Starship X*. How do you read?"

Commander Amhurst picked up the mic. "Copy you, five by five, *Starship X*. How do you copy? Over."

After twelve minutes, the radio sparked to life again. "ISS, this is *Starship X*, copy you two by two. Over."

Commander Amhurst said, "Good to hear your voice again, Steven. Over."

Twelve minutes later, they heard from Steven. "Yeah, good to hear your voice again, too, Charlie. We were wondering how bad the situation is there and if there's anything we can do for you from thirty-five million miles away."

"Good question, Steven. That's one we've been pondering here as well, and we decided as a group that you can. We could use your opinion and insight. We don't have a lot of options here. We do have a good atmosphere in the

166

Unity node, and it looks like she's holding pressure. We only turn the radio on fifteen to twenty minutes prior to our expected conversations to preserve the batteries. The Soyuz seems to have good integrity as well, but we may have depleted a good portion of the batteries sitting in it for five days with the radio on most of the time. We are a little concerned as to their ability to fire up the parachutes on reentry. If we want to do a space-walk and inspect the rest of the station from the outside, we will export a decent amount of our useable atmosphere. We are afraid to open any of the hatches into the other nodes and lose the atmosphere that way as well." Commander Amhurst looked at his crew for any input. "Do you guys have any other questions for *Starship X*? They just shook their heads, so Commander Amhurst finished the transmission. "Over."

Another twelve minutes later, they heard Steven on the radio again. "ISS, this is *Starship X*. I have made detailed notes of your situation. We've had numerous conversations about the loss of communication with CAPCOM, and we were wondering if it could be the Kessler Effect. I assume you're familiar with it. We have a theory that all of the space debris could be creating a static charge, blocking or absorbing any radio signal. We were curious as to your thoughts. Over."

167

Commander Amhurst looked at his crew, and Giulia and Henry shrugged. Commander Amhurst responded, "*Starship X*, this ISS. Until now, we haven't thought about the communication loss being related to the satellite strike, but it makes sense. Over."

"ISS, this is *Starship X*. I will discuss your situation here with the crew and get back to you at twelve Zulu tomorrow. By the way, do you know if your solar panels are producing? It sounds like you took some pretty hard hits, but I wonder if you have any power coming into your breaker box. Talk to you tomorrow. Over."

Commander Amhurst picked up the mic again. "I doubt if we have any solar panels left at all but will check it out. Roger next communication at twelve Zulu tomorrow. Over." Commander Amhurst looked at Giulia and Henry. "So, that's that for today."

They didn't know what to say, neither did he. Commander Amhurst was more aware than ever of the importance of these daily "chats" with *Starship X*. He wanted to come up with some words of encouragement for his crew or some cheerful takeaway. He really liked both Henry and Giulia and could see on their faces that another twenty-four hours before they got feedback was a long time. He loved his crew and was concerned for their future.

Commander Amhurst smiled. "Any thoughts about the Kessler Effect and the satellite strike related to radio communication with CAPCOM?"

"Well, I guess it gives us something to think about," Henry said. "What do you think, Giulia?"

"I suppose it is possible. I hadn't really given it much consideration."

Commander Amhurst nodded. "Yeah, I was thinking about it. It never occurred to me that it could be what Steven said—a static charge built up from all of the bits and pieces floating around. I guess we better find out if we have any solar cell production. Anyone know how to figure that one out?"

Henry rubbed his six-day beard while Giulia scowled in consideration of their options.

"Who's the best electrical engineer here? You guys jury-rigged the radio together."

Giulia glanced at Commander Amhurst, then at Henry.

"I guess that's me," Henry said. "Okay, let me check the breakers and see if I can find any power."

"I'll give you a hand." Commander Amhurst put the mic back and turned off the radio.

Within an hour, they knew that they had 114 volts going into the breaker box from somewhere, which wasn't coming from the Unity node's batteries. Maybe things were looking up. With a source of power available, they began searching the storage locker for a battery charger or power meter.

Commander Amhurst thought, *Well, if you play your cards right, you might just get to play another day.*

23. MARS - SOL 7 - I Will Love You

All the Way to Mars and Back

Commander Carrigan was thrilled with Carl and his team's work. He wasn't going to let those efforts go unrewarded. Not only did Carl and his crew set up the MOXIE, but they set up the Kilopower generator, started it up, and plugged it into *Staten*. Oxygen was being pumped into *Staten's* O_2 tanks at a rate faster than it was being consumed, at least for the moment.

It was SOL 7, or the seventh day on Mars. The crew had been working their normal shifts during the six-month journey to the red planet. Every ten days they were required to schedule a day off or else it would automatically appear on the roster during the long trip. Most of the crew didn't want days off during the trip but were required to take them for mental health. However, most of the crew had this day off. It was a rather unusual event to have most of the crew off at the same time. A skeleton crew was operating the ship not assigned to any diligent work other than the normal monitoring of systems.

Steven Henderson had another radio conversation with the ISS crew. The rest of the crew could be found in their quarters reading, in the game room, or in the galley relaxing and talking.

Commander Carrigan was in his office thinking and planning the next two weeks of EVAs, with the next three days entered into the scheduling system. The next big job for his crew was to go into Starship IV and remove all of the supplies, some of which they could put to use right away—like the solar panels. Most of it needed to be stocked and protected from the Martian dust. He estimated it would take the better part of two days to unload, and he would need Carl for the removal of the stringers, or the aluminum frame used to secure the cargo.

Carl was becoming worth his weight in gold. Of course, all the crew members were extremely valuable in different aspects. Commander Carrigan made a point to check on Carl and make sure he was taking it easy. He needed to be rested and refreshed for the upcoming EVAs. They had a crew meeting the night before, and *Staten*'s crew received the news about the situation aboard the ISS. They asked that their thoughts and prayers be sent to the ISS and Russian crews.

Some of the crew were visually upset knowing the options were extremely limited.

Again, the crew asked why *Staten* could talk to the ISS and not to CAPCOM or NASA. Commander Carrigan asked Steven if he could field the question, and he thought he handled it well. Steven explained that the ISS was informed of a satellite strike at approximately the same time Staten had lost comms with CAPCOM. He discussed that it may be possible that a static charge buildup from the debris may be absorbing any electronic signal, causing a comms blackout. But no one knew anything for sure, this was just a theory. The crew also had a few questions that simply couldn't be answered, like when would they be able to communicate with CAPCOM or NASA.

Commander Carrigan was satisfied with his decision to remove the stringers from the stricken Starship IV. He was sure they could be manufactured into an excellent frame for the scaffold, and Carl agreed. It was going to be a lot of work cutting all of the weld points, and they had to be careful not to damage the stringers when they cut them out. Some were undoubtedly bent from the starship falling over. Carl could do it—he had no doubt. That was becoming a commonly repeated phrase – Carl could do it. There was also some aluminum framing not a part of the stringer system that he could use to reinforce the scaffold. Carl had gotten back to him on his request about the rope and the amounts stored on board *Staten* and Starship IV. There was plenty to do the job.

Carl's design for the scaffolding was impressive and detailed. It included bolting it together at different lengths, so it could be dismantled and moved to the next starship.

Andrea was doing pull-ups while Pedro was lifting weights in the exercise room. Pull-ups are crazy easy on Mars, so instead of the normal thirty she would do on Earth, she wanted to do one hundred. Every kilogram taken to Mars costs a ridiculous amount of money, so the weight machines are adjustable hydraulic resistance machines secured to strong points on the walls. Andrea was on her twenty-second pull-up. "I'm trying to get this situation with the radio and communication with CAPCOM through my head. Even though we can talk to ISS, we can't talk to CAPCOM because of some static charge? I'm having a hard time understanding that one."

Pedro enjoyed the fact that he could curl fifty kilograms on Mars. "Do you remember the old HAM radios they used years ago? The ones they used to hand-tune?"

Andrea was on pull-up thirty-four and still going strong. "I've seen pictures of them and other old radios on display in museums."

"Well," Pedro said, "those units required the HAM operator to tune in to the correct frequency, and even when they were on the right frequency, they would often have to hand-modulate or keep tuning it in to get the best reception. Even when they were perfectly tuned-in, the atmosphere could change, and they would start losing reception and have to tune it in again or fine-tune it. If they didn't, they would lose the person they were talking to, and all they would get was static. These days, all of our radios are auto-tuning. They know when they're losing the reception, so they automatically follow the signal. Even on good days, the atmosphere can change the signal frequency. Some days the atmosphere would allow you to 'bounce' your signal off of it like throwing a ball against a wall, or better yet, the ceiling, and you could pick up people thousands of miles away."

"So, what does that have to do with us not being able to contact CAPCOM while we can talk to the ISS?"

Pedro was into his third and final set huffing and puffing. "Wow, I can tell I haven't worked out lately. It's kinda the same thing. All of the debris circling Earth is

175

generating all kinds of static at different frequencies and interfering with both CAPCOM's signal and our signals as well. It's like trying to talk to a wall with people on the other side. They need to open a window or something, so they can hear you."

"So, it's like they're blocked from hearing us?"

"Yeah, pretty much. As far as CAPCOM and the Deep Space Network are concerned, I think their signal is being absorbed like a towel absorbs water. Our signal is so weak, it probably just gets lost." Pedro finished up with his last set and started jumping around to loosen up. He had to be careful not to jump too high and hit his head on the ceiling. "Hey, I wonder how high we could jump on a trampoline here."

Andrea was on pull-up number seventy-seven and slowing down quickly. "I think one hundred is going to be too much for me today. Like you said, I can tell we haven't kept up our routine as much as normal. I'll be lucky to do ninety," she blurted out as she counted off pull-up eighty-four. "Six more. Ninety. I'm done for today. I would think you could jump about three times as high on a trampoline as you could on Earth," she said while trying to catch her breath.

Pedro turned around to watch her. Andrea was in excellent shape, as all of the astronauts were. She had great legs, but those eyes. When she focused on him, Pedro could feel her gaze touch him with a deep warmth that remained even after she had gone. "Yeah, but you weigh less here, too, so when you come back down, you don't get as much spring to go back up. I guess we'll never know until someone brings a trampoline."

A poster of the first Falcon Heavy launch when Elon Musk launched his Tesla Roadster into space caught Pedro's eye. When he turned around, Andrea was right behind him. His gaze met hers for a second too long. Instead of moving, they stared into each other's eyes as the sweat ran down her face.

"What are you looking at?" she said.

They were both covered in sweat, and he could feel her breath on his face. "I was looking into your beautiful eyes, but now I'm looking at your lips."

"Me, too."

He kissed her mouth. It was soft and warm, receiving and welcoming as she kissed him back.

As they withdrew, Andrea said, "Come with me."

In the shower, Pedro followed her movements, wiping down her toned, supple body with a damp towel.

"I'm in bunk 39A," she said. "Meet me there in two minutes."

Soon after, Andrea opened the door to her bunk, and Pedro walked in. She was already naked again. "Now," she said. "Where were we?" She giggled a little and grabbed both sides of his head, planting another kiss on him as she heard the door shut.

Pedro looked her up and down. "My God, you are beautiful."

She just giggled again and slowly laid her body next to his. As they stared into each other's eyes, Andrea kissed him. "You know, if we do this, it cannot change our relationship outside of this room. If I'm in charge of an operation, you must carry yourself as you have so far on this journey, and if you're in charge of an operation, I must act as I have to this point. I don't want to announce that we're in a relationship to everyone on the ship, and I'm pretty sure you don't either. I don't think we want the whole crew to start treating us differently."

"I don't see why it should change any. I like you, Andrea. I've liked you for a while. I also respect your judgment. None of that will change. We're simply enjoying each other's company at the moment. We're extremely fortunate to be here in the first place, and being together right now, here with you, is only icing on the cake. I'm not going to allow this moment to affect our relationship when we're working together."

"I was hoping you'd feel the same way. We're still professionals. We're just enjoying the afternoon together. We deserve to be human, too, don't we? We deserve to protect and enjoy our humanity."

With that, Pedro kissed her, and she felt her body melt into his. A deep, stirring warmth rose even more as she felt his strong quads and abdomen.

He was as aroused as she was, and it had been a long time for both of them.

She mounted him and felt her legs go weak, feeling nothing but the wonderful pressure of first penetration. Andrea felt a flood of moisture and had an immediate orgasm as she felt him go deeper. She had never felt such arousal, and neither had Pedro. It felt right. And being that they were on Mars and most assuredly the first humans to ever make love on Mars made it even more erotic.

They kissed deeply as their passion drove them to the inevitable finish. Pedro realized that he may have liked her a lot more than she liked him, but that was okay.

Andrea wished the moment would last forever as they came together in a final thrust of exhilaration and collapsed into each other's arms. "Can we do this forever?" she whispered.

They both fought to catch their breaths, and before they knew it, they had both fallen asleep cuddling naked in the cool air-conditioning of the cabin.

24. Mars - Sol 8

"Today is the day!" Commander Carrigan announced over the PA. "Full-crew pre-EVA meeting."

As everyone assembled and took their seats, the commander referred to the notes on his tablet. "I hope everyone that had the day off yesterday enjoyed it and is well rested. We have a large A shift and partial B shift EVA about to commence. C shift and the remainder of B shift, you are on ready standby. I need to go over what we're going to be doing."

As the short briefing progressed, Commander Carrigan outlined the objectives of the EVA everyone had already received in their inbox the previous night. "Good. We have one six-hour EVA today. Currently, the temperature is minus forty degrees—a typical Mars summer time morning. The EVA will take place from twelve Zulu to sixteen hundred Zulu. We have a planned break from fourteen hundred Zulu until fourteen thirty Zulu. This complicated EVA will require two commands. As a result, we will be using frequencies one and two, and both frequencies, of course, will be monitored by *Staten* Command. Carl White is deemed "*Starship IV* Command" and is responsible for all activities inside Starship

IV. Outside command is 'Command Two,' just to keep everything simple. Command Two will be Pedro Lopez. We will need both of the Mars cars and the trailer. According to the inventory lists, there should be another Mars car trailer on *Starship IV*. Let's put it to use as soon as we get it unburied. Any water hose should be staged to move to *Starships I* and *II*. Let's put some tape over the ends of the water hose to keep Mars out of our water supply. Any rope that we unbury should be set aside for ease of access as well. We will stage all inventory and mark it with the location from where it was removed from *Starship IV*, so we can find it using the inventory lists." Commander Carrigan quickly scanned the room with his steel-blue eyes. "Any questions so far?"

The crew was attentive. A few nodded their heads, so he continued. "I scanned the inventory listing and the location numbers. The location position numbers have been printed out for Command Two crew to place on the supplies once they've been staged on the surface. Command Two and Starship IV Command you will receive any coordinating information required from *Staten* Command. Today *Starship IV* Command will primarily be emptying *Starship IV*. Tomorrow, they will be removing the frame used to secure the load and hopefully start building a scaffold out of it following removal. This scaffold will allow us to gain entry into the other starships. Does everyone understand what we're doing? Carl? Pedro?"

"Yes, sir," they quickly retorted.

Commander Carrigan surveyed the crew, looking for any signs of confusion. "Okay, this is going to be a tough couple of days. Let's be safe out there."

As they were suiting up, Pedro whispered in Andrea's direction, "Hundred fifty feet on a handmade scaffold."

Andrea nodded with a concerned look as she secured her helmet in place.

Commander Carrigan went to the helm and sat down with Steven and Mary, knowing this was going to be a confusing evolution. Carl filled the Mars car trailer with much of the tools they were going to be using over the next two days, which he had carefully staged at the end of the last EVA.

They had to drive quite a bit slower to keep the tools from bouncing out, but soon enough, they arrived and began off-loading the inventory of supplies from *Starship IV*. They then sent the Mars car back to *Staten* to pick up the rest of the crew. The first thing Carl did was locate the rope he had seen on the inventory list. Most of the supplies had been shrink-wrapped on lightweight plastic pallets to help hold them together during the trip yet keep the load as light as possible. Cargo webbing had been used to secure the shrink-wrapped pallets in place, much like how airlines secure cargo to keep it from shifting during flight. After removing the webbing, Carl wrapped the rope he had previously unburied around the pallet and pulled it to the open hatch. They then picked the pallet up and slid it down the ladder using the rope. As it arrived at the bottom of the ladder, Pedro and Andrea loaded it into the Mars car trailer and dropped it off for Command Two crew, who then placed it in the correct staging position with its appropriate location label. Command Two had drawn a giant starship in the sand and were using that as a design aid to help organize all of the supplies around.

Even with the reduced gravity on Mars, it was a lot of heavy lifting. Steven was running *Staten* Command while Mary monitored and took notes for quick reference. Carl, acting as *Starship IV* Command, would tell *Staten* Command every time they pulled supplies from the next storage position. Mary would make a note of it, and Steven would raise Pedro, acting as Command 2, and let him know when to change to a new location position number. Pedro was able to put the location position numbers inside the first wrap of the shrink-wrap to hold it in place. It worked better than the duct tape they had brought with them and was easier. He joked, "At least we don't have to worry about rain as they were staging the supplies."

The *Starship IV* crew consisted of mission specialists Haratu Suzuki, Adnan Ashari, and James Galway. Command 2's crew, with Pedro in command, consisted of astrobiologist Barbara Black, Andrea Tripp, and Wayne Peters.

Two hours had gone by, and they were about 30% into the load. It was going better than they had expected, and it looked like they might possible empty it all out today.

As Carl helped James pull a particularly heavy load of shrink-wrapped hardware, he said, "This load must have been around five hundred pounds on Earth. It's almost two hundred pounds on Mars."

They put the rope through the gaps in the pallet underneath, then ran them over the top and tied them off. Every pallet had to be turned over because they were all on their sides, as was the ship. They were hauling it over to the ladder sliding it across the "floor" to the ladder, Carl said, "Just hold it there on the ladder. I'll give you a hand lowering it down."

They just finished placing it on the edge of the ladder pallet, side down, when some of the shrink-wrap caught on the upper legs of the ladder. While James was trying to loosen the plastic wrap and hold the pallet in place at the same time, the intense weight suddenly broke the plastic wrap off of the ladder leg, and it started heading down the ladder in an uncontrolled fall. James knew Andrea and Pedro were standing at the base of the ladder, and he still was holding the line wrapped around the bundle.

As he tightened his grip on the load to keep it from falling, it jerked him forward over the ladder, pulling him off balance. At the last second, James pushed with his feet, so he wouldn't get tangled up in the ladder. Carl grabbed for the rope as it whipped around between his legs and jerked away from him. Both James and the heavily loaded pallet careened straight down toward Pedro.

James was in a head-first dive over the top of the falling pallet. The loose end of the rope wound its way around the leg of the ladder and jerked it, pulling the ladder over sideways. With a loud crash, the ladder hit the open hatch door of the starship.

Andrea shouted, "Look out!" which sounded more like a scream over the radio, along with a lot of commotion and banging. It sounded like a major catastrophe over the radio.

She jumped clear of the falling James and equipment, but Pedro kept his ground trying to break James's fall—which he did. The issue with James's fall wasn't really the speed of his fall but the momentum. Unfortunately, the top of James's helmet plowed into Pedro's face shield and helmet, and both James and the two-hundred-pound bundle, pallet and all landed on Pedro, then broke apart.

It ripped open, strewing its heavy metal tools and equipment all over the ground and half burying Pedro. Immediately, compressed air rushed out of Pedro's suit as his regulator went into full free flow. James tried to untangle himself and stand up.

Over her helmet mic, Andrea announced, "Pedro is down. Pedro is down!"

Knowing Pedro must be badly hurt, Mary paged Commander Carrigan on the ship-wide PA.

Commander Carrigan, in the dayroom right below the helm, had heard the commotion and informed the crew on ready standby to suit up, then he hailed Tammy Spencer, the flight surgeon, to get medical sector ready for a possible trauma incident.

25. Johnson Space Center—Day 8

Brad Brown was on the landline with General Latimore's secretary again. Twice before, she had taken his message that he needed to speak with General Latimore as soon as possible, and each time she promised to relay the message.

This time she said she had spoken to him directly and knew he had it on his agenda to return Brad's call. She asked as to what it pertained so the general could be prepared for the call. Brad explained that it was in reference to the rescue capsule being assembled for the ISS. She understood and promised to pass the information along to the general.

Wow, Brad thought, red tape was difficult enough to get through when the satellites were working.

He had gone home the previous night to get some sleep. Now that the power was on around most of the country, the aftermath of the satellite loss was being broadcast nationwide on affiliates that still had the old-fashioned antenna transmitters. The economies of the United States, Canada, and all of Europe were a disaster. He didn't know about any other countries. It was difficult to find a station showing anything besides food trucks, helicopters, and the National Guard distributing water and patrolling the curfew.

Some personalities were arguing their theories on who destroyed the satellites, but they had no idea and no facts to back up their assertions.

Obtaining food was still a major issue because no one had access to cash. Even if they did, the grocery stores were all still mostly empty, and the frozen food warehouses had to throw all of their food out after losing power for a significant number of days. They could last a day or two on the back-up generators, but most of them had run out of diesel.

People were taking spoiled food from overflowing dumpsters to take home and eat. The army reserves were convoying food deliveries from warehouses to major metropolitan centers. Some people were going mad attacking the trucks which were trying to get supplies to the grocery stores. If it wasn't accompanied by the gun-toting National Guard, it was in danger of not making it to its final destination.

Brad thought, are most people normally insane, or does it just take a little stress?

The National Guard passed out MREs (Meals Ready to Eat), which had been produced in the millions for the armed forces. The television told people to eat nothing but canned foods and to throw out all perishable frozen foods.

Diabetics were being told not to use their insulin if they hadn't been able to keep it cold. It was still a mess all over the United States, but the situation was slowly getting better.

The president made national appearances on a daily basis, trying to explain what the government was doing to help the situation. He reiterated that martial law was still in effect, and the curfew was now at sunset since most clocks were no longer correct after the loss of the satellites and power.

The plight of the Mars Mission I crew and the ISS was quickly becoming one of the prime-time news stories, and news trucks were starting to stack up at the entrance to the Johnson Space Center. It had become obvious that there was nothing more the MMI team could do other than stay diligent on the communications equipment.

The communication attempts were still ongoing every hour, on the hour, on the same frequencies and on the back up frequencies, with no response—as usual. The ISS Ground Control was having the same issues trying to communicate with the ISS, though at least they had a telescope watching it. Not that there was much the telescope could do other than it was still there and so was the one Soyuz capsule.

It had been eight days since *Starship X* landed on Mars. The MMI team had gone over all of the figures, and they estimated that *Starship X* would soon need water and oxygen. Fortunately, they had a MOXIE and the portable Kilopower unit on board *Starship X*, which would help provide power and oxygen. Who knew if the crew on *Starship X* would be able to access the fallen *Starship IV* or any of its supplies?

Brad was sure they wouldn't be able to access any more of the badly needed supplies from the other starships because the access hatch was 150 feet off the ground. The MMI engineers had been working nonstop on the situational possibilities of what the *Starship X* crew may be experiencing and how they may be using their in-situ resources. The engineers decided that the crew had probably started up the MOXIE and the Kilopower unit by now. This would give them oxygen before the water crisis would arise. The crew on Mars would have no way of refilling their water tanks without the supplies on the other starships.

Brad's phone line rang, and he answered. "Brad Brown."

"Brad, this is General Latimore. You called?"

"Yes, General. How are things with you?" Brad cringed at his own question.

"Starting to get some things back together. Badly wish we had cell phones. I can't travel and stay in touch with my secretary. We've opted for handheld radios, and some vehicles are being refitted with mobile radio units. The old radio infrastructure and antennas all but gone now, so our range is dramatically reduced. The messages build up, and when I get back to my office, it takes me a whole day to answer all of the questions and coordinate with everybody. How can I help you, Brad?"

"I was wondering how the rescue mission for ISS is going. I need to somehow get a message on the capsule for the ISS astronauts before they board it to come back."

"A message for the astronauts before they return to Earth?"

"We have a theory that the ISS crew may be able to talk with the MMI crew on Mars. If the ISS crew are in communication via their radio with Starship X, we have an extremely important message that could be of lifesaving importance that needs to be communicated to them before they return on the rescue capsule. That is assuming you're still planning on the rescue mission."

"So, Brad, you think the ISS may be able to communicate with the MMI crew via their radio?"

"We're not sure of anything, but we think it's entirely possible. They could be far enough outside of the static barrier encircling the Earth to allow them to have radio contact."

"I see. Okay, we have been working on what is, in effect, a giant, three-layer shield covering the top three-quarters of the Crew module. It's made out of a specially designed, three-inch-thick hardened steel that can take some serious impacts and will be the base layer of the shield. Then we will be placing two layers of a Kevlar-like material with dead-air space between—you're probably familiar with it. I have been informed that a variation on the theme has been used on the ISS in different places."

"Yes, I am familiar," Brad said. "That three-inch-thick steel cover has got to be crazy heavy. It's a good thing we have an extra starship booster available for that kind of weight. When do you foresee a launch date?"

"Fortunately, we had an available crew capsule as a backup, and we're fitting the last of the protective Kevlar jacket on it now. Kennedy Space Center has a tentative launch date scheduled in about ten days. I am satisfied that we're moving as fast as possible. This is an extremely rushed launch time by anyone's parameters. I have to reiterate that this is not being advertised or released to the press, or anyone, including the families of the astronauts because we don't know if they're even alive. Not to mention what kinds of conditions they might be experiencing. So, we felt it would be best not to have it covered by the press. Kennedy Space Center is closed to the public right now under a national security lockdown. This is a national security operation."

"Of course. This conversation is on a need-to-know basis only. If it launches in ten days, that will mean the astronauts aboard the ISS will have been up there in who knows what kinds of conditions for eighteen days since the attack."

"Exactly. Do you know how you're going to get your message to the astronauts on the ISS?"

Brad had been thinking about that exact question.

General Latimore continued, "The whole operation is going to be automated. We will program the radio on the crew capsule to automatically hail the Space Station to announce its arrival. Once they board and lock the crew capsule hatch closed, it will autonomously return, requiring no input. The best thing would be something that they simply could not miss. We could program a message into the control computer that they must acknowledge prior to departure."

"Yeah, that would be perfect and If we attach an envelope with the detailed message to each of the seat belts in a way that they couldn't be removed, fall, or float away without the seat belts being undone, they will definitely see it before leaving. If we print 'Do Not Disembark Without Reading' in large print on the envelopes, they will get the message. That sounds like a good plan."

General Latimore paused for a moment then asked inquisitively, "What exactly is the message for *Starship X*?"

"Remember when we were in conference room G, I told you that the crew on Mars can't even get to their supplies because we're unable to send the authorization codes to the other starships that will turn them on? Well, we found a way for them to start up the computers, which will give them access to their supplies. They still have to access the one hundred and fifty-foot-high hatch, but it will significantly shorten the time and effort it will take to off load their supplies. This may mean the difference between twenty-one astronauts dying and a successful first mission to the red planet."

"Okay, you put your envelopes together Brad, and we'll be sure to get them to the ISS crew. I'll make sure they confirm that they transmitted the message to MMI prior to departure. Let's just hope you're right and they're in communication with the crew on Mars."

"Yeah," Brad said, "and let's just hope the heavily reinforced capsule can survive the round-trip journey, and that the astronauts on the ISS are still alive. Thank you for your help, General."

"You're welcome, Brad. We want to do everything we can to help those guys."

Brad hung up the telephone. He began to think, Well, at least we have done something for the crew on Mars. Just how the hell are they going to gain access to those hatches one hundred and fifty feet high? With an aching heart he picked up the telephone again.

26. ISS - Day 9

Commander Amhurst was doing his daily rounds of checking the atmospheric gauges. The O_2 levels had dropped a bit again, now at 17.5%. It was time to add some more oxygen to their atmosphere.

They must have had some small leaks behind all of the equipment mounted on the walls of the node. He didn't know how long the air and oxygen bottles would last. No one did. Their atmosphere would probably be the deciding factor as to when they had to leave. There were no pressure gauges on the bottles; they were just used until they were empty, and things like pressure gauges had a tendency to get bumped and broken off while in transport. The last thing they needed was an out-of-control oxygen bottle flying around inside the station or in the resupply ship.

NASA had engineers and automated programs that signaled when they would need new air tanks. Opening the tanks would wake up the heartiest of sleepers, so he went through the food inventory. Even though it was still cold in the Unity node, they were extremely lucky to have access to the food stores. He decided to eat a few dehydrated strawberries that came in a plastic pack.

Giulia and Henry started to wake up and float out of their sleeping bags.

"Good morning, sleeping beauties." Commander Amhurst quipped.

"Morning," was the mutual response.

"You guys want some nice cold coffee?" Commander Amhurst joked.

Giulia responded with a smile and a stretch. "Sounds great!"

Commander Amhurst could always count on Giulia to lighten the mood and he considered them lucky to have her with them.

"Sure," Henry responded as he began to brush his teeth. "Love some."

"I'm rehydrating some strawberries right now," Commander Amhurst said as he pushed cold water into the packet. "We also have peaches we can rehydrate, one fresh apple—well, it used to be fresh—and some cold scrambled eggs for breakfast. And of course, granola bars. By my estimate, we have a nice variety of foods to last at least another thirty days. After that, it'll be granola bars, nuts, coffee, and water." As he made his way to the dry food stores, he took out three packets of instant coffee and began filling them with water. "I was waiting for you guys to wake up before adding some oxygen to the atmosphere from the O_2 tanks. It's around seventeen and a half."

"No wonder I feel so lethargic," Henry muttered as he swallowed his toothpaste and floated over to get some water.

"Yeah, get some coffee. I have the fruit hydrating now. Whoever wants the apple can have it. I'm going to open the oxygen bottles." When Commander Amhurst opened the oxygen tanks, he could tell they were a little lower pressure than they had been before when the air blasted out. After ten seconds, he closed them and made his way back to the atmospheric gauges. They were reading 19.5 percent at 13 psi, so he decided to let the air mix and check it again in a little while. Then he thought about the power they'd found at the fuse box. "Hey, Henry! Do you think we could run that power from the fuse box into the air handler with the CO_2 scrubbers?"

"Let me check the air handlers and see what kind of power it is designed to handle. Yeah, they're one hundred twenty-volt, ten amp. If that power has much amperage, it should work. It should move some air, anyway. It's just a fan, right?"

"Yeah, I think so. Let's get some wire after breakfast and see what we can put together."

During breakfast, the usual conversation came up. When would they be able to head back to Earth? When will it be safe enough to try? What about the batteries in the Soyuz? They considered different theories on how long it would take for the debris to clear up. Commander Amhurst didn't want to explain to his crew that the debris problem could, most likely, take years. Each day they stayed in what was left of the ISS was another day they stayed alive.

The crew was all too aware that they were consuming tangible, irreplaceable resources. Once the oxygen started to fade, their time would be limited. They had to discuss these issues before they were out of time.

With breakfast finished, Commander Amhurst helped himself to another cup of coffee. He decided it was time to review the decision-making diagram. "Okay, team. I've decided that we should review some things. Since we probably can't depend on NASA to save our butts here, this is what will come to pass in the next unknown number of days. I'd like to discuss and consider our options so we're all on the same page and prepared when we take the actions that could easily make the difference between life and death for us. From my perspective, oxygen is our most critical issue. Starting the air handler with the carbon dioxide scrubbers gives us some critical breathing room. Sorry for the pun."

Giulia and Henry smiled.

"I believe the oxygen will be our limiting factor, so we need to consider our options before it gets to critical levels. I've been thinking about the rest of the Space Station quite a bit. As I see it, we have two possible choices. First, if I go out for a space-walk to evaluate the rest of the Space Station and the solar arrays from the outside, we may be able to reclaim some of the arrays and evaluate the rest of the nodes. It will allow a larger perspective of the condition of the entire station, including the solar arrays. It may be possible to reclaim, for example, the Russian Zvezda node, where we can manufacture oxygen from the solid fuel oxygen generator.

"What we're talking about here is taking some risks. Assuming that one day soon we're down to the last of the oxygen and can't find any other sources, we'll have to make one of a couple choices. And please chime in if I'm wrong or you see another alternative. We have to make these decisions before the oxygen levels get so low that we can't think clearly. I think I should be the one on the space walk, just because I'm in command, and for no other reason. We are all qualified. As we've discussed in the past, that's one of our options. A second option is all of us suit up and go into the Zarya node, then shut the hatch as fast as possible to try to preserve what is left our atmosphere in here. Which won't be much. Then we'll conduct a thorough evaluation of both the Zarya and the Zvezda nodes from the inside. That will give us a more comprehensive picture of the two nodes to see if they're salvageable. We can start the salvage operation, and maybe even get the solid fuel oxygen generator running, then start to repressurize the other two nodes. Of course, this action risks the atmosphere in here, so we probably shouldn't pursue that until our atmosphere in here is pretty much spent. The problem with that is if we can't reclaim the two nodes and start the solid oxygen generator, then we're stuck in our space suits. Period. I guess at that point, we'll climb back into the Soyuz, head for home, and hope for the best."

"That sounds pretty scary," Giulia said. "It's like putting all of our cards on the table at the last minute."

"Once we go into the Zarya node, there's no returning. We are committed. Either we make it work, or we leave and take our chances in the Soyuz."

"Okay, I don't think I care about waiting for the last minute. If we do the space-walk option, we only lose the atmosphere during the equalization, not the whole Unity node."

"Yeah, I'm guessing we might lose maybe one day of breathable air when I do the space walk, but I should be able to tell from the outside if either node is damaged beyond repair."

Giulia sipped on her coffee, letting it go to free her hands. As she began moving her hands all about, the guys knew she wanted to say something, but she still had the straw in her mouth. "If it looks intact from the outside, we can determine that the Zarya and Zvezda nodes may be intact enough to reclaim. Then we can take our chances on the other two nodes if it looks like they may be intact. If not, we still have another couple of days of good air in here—hopefully. We could regroup and reconsider our options then?"

Commander Amhurst and Henry both agreed.

"Yes, I wouldn't even get out of my space suit if it looks intact. We could wait until we're down to the last two or three days of breathable air, then I could go out and do an inspection on the outside. If it's damaged beyond recovering, we'll know not to try making an entry into the Zarya node, and that might leave us at least a couple days of air."

Giulia put on a sly grin. "See? You smart guys still need the occasional direction from the Italian sector." She threw her head back in laughter.

It was the first time in a long time anyone had truly laughed on the Space Station, and when they were making life and death decisions, a little laughter went a long way.

"So, it's settled then," Commander Amhurst said. "We will wait until we think we're on the last few days of breathable air, then I am going to take a walk. Unless, of course, we discover some new information in the meantime."

"Sounds like that makes the most sense at the moment," Henry agreed. "Now, let's take a look at that air handler."

27. Mars - Sol 9

As soon as Andrea got back on her feet, she helped the stunned James Galway get up as well. They hurriedly unburied Pedro. His helmet lights were blinking on and off, and his regulator was making a shrill screaming sound. The sound of the free-flowing regulator was loud enough for everyone to hear above the depressurization alarms.

Finding the hole in Pedro's suit was priority number one. Andrea pulled out her roll of duct tape and started looking for penetrations as she instructed Wayne and Barbara over the radio to bring the Mars car. She found a couple of small tears and penetrations around Pedro's quads and knees. James straightened Pedro's suit over each hole then Andrea covered each of the holes with the strong tape, wrapping the holes with three or four pieces of duct tape. Pedro's regulator slowly started returning to a normal volume as his suit returned to a more humane atmospheric pressure and the low-pressure alarms quieted down as well going off one at a time.

Wayne and Barbara Black brought the Mars car over. Wayne picked up the ladder and replaced it on *Starship IV*, so Carl's crew could get down. They put Pedro across the back seats and sped off to *Staten*. On the way, Andrea informed Steven via her helmet mic what had happened as Pedro mumbled incoherently.

Upon arrival back at *Staten*, he was immediately loaded onto the deck of the crane as Andrea blew him off with the high-pressure hose. He was brought into the equalization room, where Commander Carrigan and the ready standby crew had just equalized to the Martian atmosphere.

Before long, Pedro was out of his Mars suit and in medical, being evaluated by flight surgeon Tammy Spencer. Her immediate head-to-toe trauma assessment revealed that Pedro had a hyperextended right knee, a small contusion to his forehead, and a possible concussion, along with some smaller bruises and contusions. Neurologically, he was aware and responsive, but he couldn't remember what happened. He was given a local anesthetic for his painful knee, which was already red and swelling. With a little manipulation, Tammy was able to get his knee back in the correct position, and he was left to rest in medical.

After thirty minutes, all vitals were within normal limits, and his pain was dissipating. There didn't appear to be any rapid decompression illness or barotrauma, lung sounds were clear and equal, and his oxygen saturation was 98% on room air. Tammy placed him on oxygen as a precaution. She said she wanted to keep him for at least twenty-four hours for observation.

She left medical to get Andrea's full story about what happened, then told Commander Carrigan she felt Pedro was extremely lucky. She said he'd be walking, albeit gingerly, with a leg brace and crutches in a few days. After some rehab, assuming there were no medical complications, he could probably return to light duty in thirty days. She hoped there would be no long-term tendon or ligament damage that might require surgery.

Once Carl's crew from *Starship IV* arrived back at *Staten*, Commander Carrigan organized the entire crew. James did most of the talking at first, describing what happened. Then Carl explained how he saw the accident unroll. As a crew, they went through a whole post-accident briefing and evaluated every aspect.

The common conclusion was that they were performing a difficult operation without the necessary equipment. The second point of consideration was whether they needed to off-load the heavy load as one piece or could they have broken it down. The group decided they could've taken the load apart. It would've taken significantly more time, but James would not have been pulled over the ladder and dropped the load if it had been broken down into smaller and lighter pieces.

After the review, Commander Carrigan sent James to medical for a quick assessment after he discovered that James had fallen the twenty feet onto Pedro. Forty-five minutes later, Tammy cleared him for light-duty work after twenty-four hours of rest. James said he was fine, thanks to Pedro blocking his fall.

Commander Carrigan put C shift in charge for the remainder of the shift and went to his office to fill out the accident report forms. He was grateful that neither Pedro nor James were more badly injured. After filling out the report, he went to the helm and asked Steven to join him for an EVA back to *Starship IV*. He asked Mary to do the comms. He wanted to take a look at the scene in person.

Upon arrival, the scene was exactly as everyone had reported. Commander Carrigan and Steven looked at the ladder set-up and couldn't see any way to secure the ladder without causing a tripping hazard or significantly complicate the off-loading operation. It just wasn't tall enough to do the job. They considered placing rocks around the foot of the ladder to help secure it in place since they didn't have any other way to secure it. They decided that the crew could probably make some sand bags out of some of the supplies or extra shrink-wrap which could secure the feet of the ladder.

They discussed the set-up and if there was anything that could've been done differently. This wasn't a "normal" operation, but then again, *Starship IV* wasn't expected to fall over and have to be manually off-loaded. In the end, they considered that maybe the time frame to accomplish the task had been too short and caused the crew to rush and become too anxious to get the job done.

Considering that the crew was ahead of schedule, it could hardly be the time frame. They weren't rushing due to time constraints. Still, if Commander Carrigan had been on the scene prior to the accident, he may have been able to recognize the danger. *Starship IV's* command, crew, and even the ground crew should've been able to recognize the precarious instability of the ladder. Commander Carrigan was disappointed with himself for not being there observing or maybe acting as a Safety Officer.

It was hard to find fault with Pedro, who may have been holding the ladder or standing on the bottom rung to stabilize it. After all, he broke James's fall. It looked to Commander Carrigan that James may have suffered a broken neck if the pallet landed on him while he was upside down at the base of the ladder. If Pedro hadn't held his ground and blocked James's fall, who knows what the outcome might've been.

Commander Carrigan shook his head as he turned to Steven. "It looks like it's time to award Pedro the medal for Valiant Performance."

"Yeah, I guess, though I've never heard of astronauts getting medals before."

"Well, if it weren't for Pedro, it might've been much worse."

"I agree."

28. Mars - Sol 10

The next day, in a full-crew meeting, Pedro was awarded the "Valiant Performance" citation, which was as close to a medal as Commander Carrigan could find. Pedro was also rewarded with a standing ovation from the rest of the crew, much to his embarrassment.

The crew suited up for another EVA to get as much of the job done as the job would allow. Commander Carrigan suited up as well and acted as a safety officer on the scene. There was nothing he could recommend differently, so after a few hours, he walked the mile back to *Staten* by himself—really alone for the first time in months. He had forgotten how nice it was to be alone, and the walk picked up his spirits. He realized that he was actually grateful that no one was permanently hurt.

Surprisingly quickly, the crew finished emptying *Starship IV* with no further incidents. Some nylon bags had been found and filled with Mars soil to stabilize the ladder. The speed of the accomplishment astonished Commander Carrigan. He didn't know if he should praise them for their efforts or chastise them for being too efficient and effective. When they saw his confusion, they reported to him that most of the remaining inventory on *Starship IV* was lightweight, which included the Mars car trailer, and they had already attached it to the second Mars car (called MC2) and put it to good use.

James Galway didn't seem to have any significant injury and spent time with Pedro. Pedro was confused about the attention since he couldn't remember the accident or his part in saving James. All he could remember were incredibly beautiful hazel eyes, which brought a warm, comfortable feeling with them. This he kept to himself until Andrea swung by. She stopped in to see Pedro a few times a day, and he was appreciative of that attention, but it was different in medical: they were or could be easily observed.

29. Mars - Sol 11

It was time to remove the framing from *Starship IV* and then weld the scaffolding together. Commander Carrigan was having second, third, and fourth thoughts about a 150-foot scaffolding after a crew member just fell 20 feet down a ladder. He had engaged Carl in conversation on more than one occasion to find another way to open the hatches on the other starships. They just didn't know of any other way, and the supplies on all of the starships had to be off-loaded sooner rather than later. They needed the water lines to refill *Staten's* water tanks as soon as possible. This was not a convenience; this was a requirement.

"Ah, Carl ... just the man I wanted to see," Commander Carrigan said. "Take a seat, sir. I'm nervous about the scaffold since the accident with the ladder on *Starship IV*. We really don't want a repeat performance, and this time we are talking about one hundred fifty feet up—not twenty."

Carl sat down and pulled a file out of his briefcase, rubbing his chin as he examined a drawing. "I'm sure I can put together a stable scaffolding using the rope to secure it to the Starship every twenty or thirty feet. It will use up most of the rope we took off of *Staten* and *Starship IV*."

"I want to personally inspect it before we implement it. Not that I don't trust you, Carl. It's my responsibility." Commander Carrigan was still feeling a good bit of remorse about the accident and didn't want to repeat the situation.

"Of course, Commander. You've seen the design, but that's not the same as seeing the finished work. It is going to take two or three days, no longer."

"Carl, I know you're loaded down with all of these responsibilities. Is there anything I can help you with or give you more support for?"

"The crew of four you assigned are fine for the job. I don't need any more people to assemble the scaffold. I don't want them using the cutter though. I'll do the cutting. It's too important to keep the framing intact. The other three can be off-loading and staging the stringers and framing as I remove them. It won't take long to remove them, and then we start to weld. One, maybe two people can hold it in place while I weld and teach them to weld. We'll trade off positions, so we can rotate around and take breaks without interrupting the work. The fourth person will retrieve another length of frame from the staging pile. It's pretty simple at that point."

"There's just one more item that I need to cover, Carl. I need to get an update on Staten's vitals from your perspective."

"I knew you'd want my take on our water and oxygen levels. The bottom line is, we're going to need water in the next two days."

"Yeah, I know. We're cutting it close. Do you know how much water line we got off *Starship IV*?"

"There was one thousand feet on *Starship IV*," Carl said, "and we have three thousand feet on *Staten*."

"Yeah, that's what is on the inventory lists. I figure we need almost six thousand feet to make it from *Starship II* to *Staten*."

"I estimated that much, as well, a little over a mile. The inventory shows three thousand feet on *Starship V*, another three thousand feet on *Starship VI*, and three thousand feet on *Starship VII*.

"Which is the best one to start off-loading first? Assuming we can gain access."

"We're going to gain access, Commander. Don't you worry about that. I suggest *Starship VII*. It has the third Mars car and trailer, which we could use. It also has a better variety of different foods which we do not have on *Staten*," he said with a sheepish grin. "As well as the water line. Let me get to it. If you want to rotate the crews out there with me, that's fine."

"So, you're telling me we have about two days of water at the current usage, and it'll probably take two days to get to the needed water line?"

"At the current usage rate, yes, that's right." With that, Carl headed off with his crew to *Starship IV*.

Pedro's injury and James Galway's fall was a grave reminder to Commander Carrigan that MMI was functioning way outside of its designed parameters. It also gave the commander reason to pause. He knew they were lucky no one was killed in this incident. If someone fell from 150 feet, they would die. If they didn't get the supplies they needed off the starships, his whole crew could die from lack of water and eventually face starvation.

They needed water. *Staten* was designed for recycling as much of the water as it possibly could. Even urine was recycled. Humidity was taken out of the air they breathed, and after sanitizing and putting it through the reverse osmosis system, it was added to the water tanks.

Every EVA wasted not only precious air; it also took the humidity with it. Commander Carrigan took Steven and Barbara Black out to inspect the work Carl and his crew were doing, as well as to inspect *Starships I* and *II*. He also had a small deviation in mind, which he thought might be interesting.

He asked Andrea to monitor the comms, and Carl was acting Command at *Starship IV*, utilizing frequency 1. Commander Carrigan told Andrea he would be Mars Command, and they would be on frequency 2 so as to not confuse the crew working on *Starship IV*. He then asked Barbara Black if she had evaluated the soil samples they took on the first day of the EVAs. She said she had and determined they did indeed contain perchlorates, reinforcing the need to decon prior to re-entry.

Barbara was constantly swabbing every nook and cranny and taking her swabs back to the lab for ongoing analysis of contamination and microbiological activity of any kind. She took a sampling kit with her everywhere she went and could be found crouched in the odd corner or swabbing areas of the galley, showers, and bathrooms. If she discovered anything of significance, she would report it to the commander immediately.

They arrived at *Starship IV* in MC2. Carl's work was going along much quicker than Commander Carrigan had expected. He already had all of the framing and almost all of the stringers removed from *Starship IV*, which was probably more than would be needed. *Starship IV* had eight sets of aluminum, L-shaped framing that ran the entire length, from top to bottom, for attaching the cargo netting and securing it in place. Carl had removed the Kilopower unit from *Staten*, which provided the power to run his welding equipment.

Commander Carrigan decided to leave Carl and his crew alone and head over to starships I and II, taking with them the 4000 feet of water line in the trailer that had just been off-loaded from Starship IV. Unfortunately, there was too much line to move all at once, and the trailer couldn't take the weight, so they had to make four trips to get it all moved over to *Starship II*.

Commander Carrigan asked Steven to drive so he could take in the Martian scenery. At seventy-four degrees and a full, noonday sun, it was a beautiful day on Mars. The red Martian soil and rocks bore a stark contrast to their white Mars car. *Starships I* and *II* gleamed in the sun.

They off-loaded the water line, staging it approximately every 1000 feet, starting at *Starship II*. A quick inspection of starships I and II revealed that all of the autonomously running equipment was working as previously verified. The water tank was full, hydrogen was being manufactured, oxygen was being made, as was methane. They almost had enough methane to refill *Staten's* fuel tanks. In another month, enough fuel would've been made for the trip back home when Earth and Mars's orbits had moved close enough together again.

Satisfied that the starships were functioning as expected, they were done with the more difficult objectives of the day's EVA. Commander Carrigan suggested they look at one of the cliff outcrops that had a dark spot located on the canyon wall around some large rocks. This outcrop had been visualized by the Mars Reconnaissance Orbiter and had been deemed a location of interest.

The charge on the Mars car was at three-quarters full, and the outcrop looked to be about 3 miles away, or maybe a 30 minute ride. Commander Carrigan estimated that they could see over 20 miles in the thin Martian atmosphere. They tried to drive as much as they could in a straight line, avoiding a boulder field and the large meteor craters.

Commander Carrigan explained to Andrea where they were going and why. She could see their beacon on *Staten's* locator map, and he didn't want her to get concerned. Twenty-five minutes later, they stood looking up at the dark spot on the canyon wall, partially obscured by some large boulders. They crawled thirty feet up the side of the stone and soft sand cliff and came around a large boulder blocking their view of the black spot they could see from the valley floor.

No autonomous rover had ever seen a cave or climbed up the side of a soft-soil cliff with this type of angle or degree of inclination. So, this was virgin territory.

Steven was the first one to see it. He looked back at Commander Carrigan. "You're gonna want to see this!"

Commander Carrigan stood next to Steven as they surveyed the entrance.

Barbara walked around the two of them and gasped, "Oh, my God! A cave!" she pretty much screamed into her helmet mic.

Andrea had to adjust the volume down on her headset back on *Staten*.

Commander Carrigan let out a hoot as he looked inside what was definitely a cave. The opening was only 6 feet wide, just under 6 feet high, and fairly round. They could see about 15 feet inside. It looked to them that the floor turned darker in color deeper into the cave, but there was also less light. They told Andrea on comms that they were going into what appeared to be a cave three miles northeast of *Starships I* and *II*, and they asked her to give them a time stamp and expect possible loss of communications with the group.

"I have you on the locator map. Time stamp is zero six twenty-one Zulu, be careful." She made a quick note at the communications desk.

"Zero six twenty-one Zulu. We will stay in contact as long as we can."

The group checked their watches and confirmed the time. Commander Carrigan set a stop watch and walked 3 feet into the opening, looking up and down with every hesitant step. "Barbara, you'll want to sample some of this soil."

Barbara walked in next to Commander Carrigan and started putting samples in little glass bottles from her case. She then told Andrea which numbered bottles she was using for the different samples, so she could review it back in the lab and have a documented record without having to write everything down.

The crew turned their helmet lights on as they progressed 50 feet into the cave. They began seeing small reflections on the ceiling, growing larger the farther they went into the cave. The further they went the larger the reflections in their lights. They were crystals.

After another 100 feet, Barbara said, "Make sure not to touch any of these crystals. We don't know what they are yet."

The crystals seemed to grow out of cracks between the rocks, along the ceiling, and on some of the walls of the cave. Some of them were up to 2 feet long. They walked in another hundred feet, and the narrow, 10-foot wide cave opened up to a larger, more spacious tunnel 30 feet wide. As they progressed, it curved around to the right as it descended.

The floor continued to angle downwards as they scuffled through the dark sand that appeared darker and more compacted the farther they went. Occasionally it would flatten out for a few feet and then start to angle down again. They couldn't see very far ahead so every step and every corner was a mystery.

Commander Carrigan said, "Make sure and watch your footing. We don't want anyone falling into a cavern or something."

"How deep do you think we are, Barbara?" Commander Carrigan asked as they continued on a slow, gradual descent and rounded another corner.

"Maybe two hundred feet below the valley floor where we left the Mars car. We have come a pretty decent distance already."

As they walked a little further along, Steven said, "I'm going back to the Mars car to get the atmosphere meter and the flashlight."

As they waited for Steven, they examined the beautiful ceiling. It shimmered in multiple colors and reflected off their helmet lights.

Barbara was busy swabbing the clear crystals and stashing her samples in her swabbing kits. "They look like quartz, or maybe salt crystals," she said in amazement, then broke a small piece off the side of the cave using the swabbing to handle it. She hailed Andrea but received no response. "Try to remember that we lost comms at zero six fifty-eight. We've been in here for almost forty minutes already."

Steven was back with the atmosphere meter and taking a reading. "Hmm," he said. "The temperature seems to have risen a bit. We're at eighty degrees, and the humidity readings are ninety-five percent. I'm picking up readings of five percent oxygen and about two percent methane. That's the highest oxygen and methane readings I've seen or ever heard of on Mars."

"Methane could be expected from geological sources, but oxygen? Oxygen wouldn't normally be expected. At least not from a geological source." Barbara whispered.

They continued their walk into the cave, continuing down another 50 feet, when the floor took a deep dive into what appeared to be a large cavern.

"Careful here, guys!" Commander Carrigan didn't want any more accidents.

At the bottom of a 50-foot drop, something dark and shiny was moving. Steven's light reflected off the bottom of the cavern like a distorted mirror onto the opposing side of the cave. They could see the cave continuing beyond the moving surface. It looked like a black pool slowly boiling up from the bottom. They couldn't hear any noise or sound.

"The atmosphere in here getting thick and it is foggy with ninety-eight percent humidity. Ninety-eight percent! This is like Florida!" Steven said.

"Ha!" Commander Carrigan laughed. "Can we get down there?"

Barbara looked around in amazement at the beauty and unexpected find. "I think we can get down over there."

Steven shined the light in the direction of a more gradual slope that led around the bubbling pool as they made their way around the inside of the large cavern. "How far across would you estimate this cavern to be?"

"I think it's maybe one hundred feet in diameter, maybe more. It's hard to tell because of the fog," Commander Carrigan replied.

They made their way to the bottom of the cavern next to the pool. As Steven continued monitoring the atmosphere. "I'm reading one hundred percent humidity at eighty-nine degrees," he said. "Not only that, but I'm reading that we have an atmospheric pressure of two psi." Something dropped onto his meter and splattered. "Hold on! Barbara, come take a sample of this!"

Barbara followed Steven's gaze and his flashlight to the ceiling 100 feet above them. It was glistening with shiny stalactites and a spongy material, like something on the beach after a storm, like seaweed hanging down in short strings. She took out a swab and sampled the drop.

Commander Carrigan was looking at the pool. "I . . . I think that's water! It certainly looks like it!"

Barbara immediately took out another glass jar and carefully removed a sterile spoon, sampling the liquid. "Andrea. Andrea. *Staten*, do you copy?"

Again, no response.

They walked around the pool and looked farther down into the cave, where it exited the large cavern then continued.

Steven shined his light where the cave continued around another corner. "What type of formation do you think this is, Barbara?"

"Off the top of my head, I'm guessing this is a lava tube. I wanna see where that leads."

"Okay, we've been out of radio range for almost thirty minutes. I suggest we head back before we make Andrea too nervous. I'm showing we've been in here for almost an hour already. Next time we bring the camera, and we organize this spelunking activity a little more with safety lines, lights and equipment. The crew isn't going to believe this. Barbara, do you want to take anymore samples of that liquid before we leave?"

"Yeah, I think it may be water. It certainly looks like it." After she took another two samples, they made their way back up the steep embankment.

Before leaving, they all turned around and had one more look.

"How deep do you think we are in here?" Barbara asked.

"Deep enough to give us just a little more than double the atmosphere on the surface and an increase of fifteen degrees as well as a completely different mixture of atmospheric elements." Steven responded in awe.

"I think we're pretty deep. I would guess we've descended over four hundred feet. That cavern floor is probably fifty feet to the pool itself. I can't wait to get a look at my samples."

Commander Carrigan was looking at the ceiling again following Stevens light. "I bet that the temperature in here is more consistent than the temperature outside. Next time we come back, Barbara, bring a thermometer. I'd like to get the temperature of that pool."

"Yeah, that would make this a much more comfortable microclimate than outside. And it doesn't receive all of the radiation the surface does. Does anyone want to guess if it is like this year round?" Barbara posed.

"That is a very good question." Commander Carrigan said.

They were all in complete awe of the cave. With that, they slowly walked out, being careful not to touch any of the crystals. After reestablishing comms with Andrea, they informed her that they had exited the cave and were making their way back to *Staten*.

Immediately upon their return, Barbara went to the lab and began carefully analyzing the liquid samples. Within a matter of minutes, she had a basic analysis of the liquid and had inspected it under the microscope. She immediately went to Commander Carrigan's office, where he was writing up notes from the excursion.

"I've got something to tell you. Water," she said, with an expression that said she was going to cry.

"What?" Commander Carrigan said. "You've got to be kidding me. Really?"

"Water!" she repeated with a giant smile. "It's water, sodium chloride, magnesium chloride, potassium chloride, and some other salts . . . and no perchlorates! But there is something else." She began to cry and then laugh. She bit her index finger with excitement. The usually scientific, demure, and serious Barbara was having a breakdown.

"Water . . . and something in it?" he almost yelled. "What is it? What is it?"

"I don't know," she said, trying to get a hold of herself. "It's going to take time to identify whatever it is." As she regained her composure, she cleared her throat and said purposefully, "It looks like a diatom... Maybe."

Commander Carrigan's face lit up. "You discovered water in liquid form, and possibly life at the same time?"

"You could drink it, I think," she added, giggling. "I haven't done a full analysis, but the PH is seven-point-nine, just slightly basic, which means it probably tastes good, too. It's definitely water. Water with electrolytes. It might actually be good for you. I wanted to give you the news as soon as I had it confirmed. I'm going back to the lab to take another look at this diatom thing, or whatever it is. I'm also curious about the soil and crystal samples I took. I . . . I can't believe it!" She turned to leave, still shaking and wiping tears from her eyes. She turned back to the commander and shook her fists in joy. Then headed back to her lab.

"Congratulations!" Commander Carrigan yelled after her as she scampered away.

Commander Carrigan couldn't focus on anything else. He had to go to the lab and see this thing Barbara had found.

At this time, Steven was making his usual radio shout out to the ISS. There was nothing new for either to report. He didn't think they needed to hear about the falling incident and the injury to Pedro. The ISS crew had nothing they could do but wait it out until their atmosphere was almost gone, and then they were going to attempt what Steven thought of as a hail Mary toss to the end zone. If it worked, they could stay for an indefinite amount of time, or at least until their water ran out. He wished them good luck, patience and Godspeed and signed off.

What kind of hell must it be knowing that sooner or later they're going to have to climb in that Soyuz again with low batteries, hoping that they can survive reentry, and trust that the parachutes open. If they aren't smashed into pieces on the way. He never thought he would question space travel. And here he is on Mars.

Soon enough, Carl and the crew working on the scaffolding returned to *Staten*. Commander Carrigan announced a full-crew meeting in thirty minutes over Staten's all stations, PA.

He had popped into Staten's laboratory to see Barbara Black previous to the meeting. She opened the lab door, which usually remained locked while she was working.

He greeted her with a big smile and surprised himself when he said, "What's up?" knowing that wasn't like him.

Barbara looked at Commander Carrigan. "Mask. Gloves."

Back to her scientific self, he thought as he followed her into the lab. "Of course, we don't want to contaminate these samples, that's for sure. We're having a crew meeting in about fifteen minutes. We need to tell the crew what we know, and probably a fair amount of what we suspect. Where is the diatom thing?"

"Diatoms," she corrected him. "They're in the incubator. Since we didn't get a temperature of the water sample I took, I'm trying to keep them at thirty-two degrees C—the same temperature as the room we found them in. If they are diatoms like on Earth, they won't live long. Most diatoms have short lives. Maybe a few days."

"What do you think they may have been utilizing as a food source?" He couldn't help asking, even though he knew better.

"That's why I'm trying to analyze the water for nutrients and anything else. I also need to go back and get more samples of the water, the edges, the bottom of that pool, and the ceiling of that cave. Most diatoms in the oceans and fresh-water sources on Earth live off of photosynthesis, but some live off of carbon sources for energy—especially the ones found in caves. I also want to leave a digital recording thermometer near that pool, so we can get the temperature changes over time."

"I need to talk to Steven before the meeting. We have to tell the crew what's going on and what it is we've discovered. So, I'll describe our little journey today, and I'll take us as far as approaching and entering the cave. Then I'll turn it over to you to explain since you're the specialist in these things. Are you comfortable declaring that we found water?"

"Sure. I feel comfortable saying we've discovered water and a living organism that so far looks very much like a diatom. They'll have questions all night if I don't tell them everything we know and everything we don't know. And even then, they'll understandably be full of questions. I'll handle it. I'll see you at the meeting. Let me put some things away," she said as she started busily cleaning up the small lab.

"See you at the meeting in ten."

"Okay, see you there," she replied with a smile. "I'm so happy to be able to put my skills to work."

"I'm sure you are. There's no one better for the job."

Commander Carrigan went to the helm where he found Steven finishing up the communications log. "So, what do you think of our discovery today?"

Steven looked up from the log screen and smiled. "I can't believe it."

"We probably discovered the first liquid water on Mars."

"It was water?" Steven said.

"Looks like it! I just wanted to give you the basics. She has verified that it is water, but that's not all . . ." Commander Carrigan said with a wide smile. "That's not all!"

At the beginning of the meeting, Commander Carrigan made the announcement that *Staten* was under strict water restrictions until the water line was completed, meaning no more showers until the water line was intact in a couple days. He asked Carl to give everyone a quick update on the construction of the scaffold, which was moving along at a rapid pace.

The team was working like a smoothly oiled machine. They were about to start adding the cross-support structures to the first two levels already. He explained that they were bolting the structure together every 25 feet, so it could be dismantled and easily moved from starship to starship. Even then, that meant six different pieces of the scaffolding had to be moved each time they moved to the next starship. The base would be the worst piece to move because it was so wide and bulky. Like a pyramid the upper structures were smaller and narrower.

Fortunately, the approximate 1/3 weight on Mars helped a lot. He had devised a way of using the Mars car and the trailer as a type of dolly with two astronauts walking alongside to keep it stable as the Mars car in front pulled it along.

Commander Carrigan asked Carl how he planned on getting the supplies down once he made entry. Commander Carrigan was a little concerned about asking this question because he himself didn't know the answer to that question. Carl was right on top of it, though, and said he would be disconnecting the computer-controlled crane and rewire the contacts to the cranes motor. He was going to rig an up and down switch utilizing the starships existing battery power.

"No big deal." Carl smiled back.

"Good to hear." Commander Carrigan felt encouraged.

Commander Carrigan explained the rest of the trip to *Starship I* and *II*, stating that they had delivered four thousand feet of water line and staged it for rapid assembly as soon as they had enough line to connect to *Staten*. He went on to explain that there happened to be one of the formations identified by MMI for exploration, an "area of interest" not far from the *Starships I* and *II*. He outlined their trip up to the point of discovering the cave, then he handed off the rest of the explanation to Barbara, who explained the whole exploration event from the minute they arrived at the cave to the discovery of the liquid pool, which she had since been able to identify as water.

The crew clapped and gasped with amazement.

Tammy Spencer asked a pertinent question. "How can water in a liquid state remain liquid and not immediately sublimate?"

Barbara had been thinking about that ever since they left the cave, as had Commander Carrigan and Steven as well. "Steven was giving us atmospheric changes from the atmospheric and environmental meter. From the moment we entered the cave, the atmosphere began to display changes. As we approached the lowest point of the cave, the atmosphere was definitely changing. The lower we went, the warmer it became, and slowly, the atmosphere began to increase in pressure until we arrived at the edge of the pool of water, where the temperature had risen about twenty degrees. And the atmospheric pressure was more than double that of the surface pressure. I believe he reported that it was at two psi." She looked at Steven, who nodded in confirmation. "We also noted the humidity was approaching one hundred percent, and it appeared to be producing—or at least we witnessed—what appeared to be a fog or a humidity high enough to perhaps condense on the walls and the ceiling . I'm guessing the added atmospheric pressure was keeping the water in a liquid state, or as it appeared, it was slowly boiling instead of sublimating directly into the atmosphere. If I didn't know better, I would've guessed that we were in a hot spring in an Icelandic lava tube in Mars suits. We obviously need to evaluate it further."

The next question was from Mary Pfeiffer. "How big was the cave?"

Commander Carrigan explained that the size was somewhat difficult to explain since it started about 30 feet up the cliff face, where it was only 6 feet in diameter and slowly opened up to over 100 feet wide after they had descended to somewhere around 400 feet below the valley floor.

"What were the crystals, and what could cause them and the stalactites?" Anan Ashanti asked.

"All I can imagine is that the processes that operate in caves on Earth are doing the same thing on Mars—with the lower gravity of Mars." Barbara stopped for a moment to think, then added, "As you know, since we've been here, the temperature in Valles Marineris has been way above freezing during the day. It's also understood that at least during the summer, Mars expresses humidity, which we are using to convert to water. I'm just guessing, but as the ground warms up, the ice in the ground melts, and it slowly percolates down through the rock and accumulates below, such as in the cave pool. Apparently, it's low enough to create its own microclimate. We will be studying this phenomenon for some time." She glanced at Commander Carrigan, who smiled and nodded again.

"Of course, this cave is fascinating." added the commander.

"The cave continues on where we stopped. There is obviously more to investigate. It appears to go deeper. Who knows what secrets it has to reveal. The final aspect of today's little journey is, when I arrived back here at *Staten*, to say I was interested in evaluating our samples would be understating it just a bit.

A few giggles and some laughter could be heard as she paused.

The liquid, as you know, turned out to be water. As any biologist would, I evaluated it under the microscope while my other tests were being run. And guess what." No one said a word. "I found tiny algae, I think. It looks like diatoms, but it is definitely life! We found life and water on Mars—all at the same time!"

There was no organization to the space ship called *Staten* for a good twenty minutes after Barbara uttered those words. There was shouting, screaming, handshakes, high fives all around, and many, many colorful curse words uttered in a rainbow of emotions. After all they were not only the first crew to land, walk and live on Mars but now they also, as a crew discovered water and apparently the first life forms on another planet!

Commander Carrigan stood back and took it all in.

Each and every member of MMI realized the significance of this moment. Man had been looking for extra-terrestrial life for years. The now historic Mars Mission One crew had just added two more firsts to the quickly building first-in-history list.

Commander Carrigan loved every minute of watching his crew enjoy the benefits of their hard work and dedication. As he watched them, he considered what each of them had done, all of the risks they had taken to get here and to assist with the success of the mission. They deserved it, and God was it gratifying for him to see them celebrating this success—this milestone of the journey. Despite all of the issues with the mission, the loss of communications with the Earth, *Starship IV* falling over, Pedro being hurt and who knows what this Kessler Effect situation may mean for them, these crew members of *Starship X* and MMI worked their asses off to get here. They were good people who made it all worthwhile and they deserved every second to celebrate the success of today's discovery. Commander Carrigan sat back watching by himself as tears came to his eyes, the dedication, the common goal, the imagination, the risks they all took in the name of discovery and here it is right in front of him. Beautiful. No man could be more appreciative of this moment. He loved his crew. He loved their dedication and he loved their heart.

30. Johnson Space Center -
Day 12 - Weather or Not?

The engineers at Cape Canaveral installed the message of acknowledgement on the crew capsule's navigational computers and programmed it to not respond until an acknowledgement had been entered. They also seat-belted in three separate envelopes. The capsule would not return without the crew answering the question, "Did you send the message to *Starship X*?" There was no way they could miss it.

The capsule would be placed on the booster and then the launch platform the following day. Fueling would begin the day after that, assuming everything went smoothly. They were hoping for launch in three days.

The hurricane season starts on July 1ˢ in the Caribbean, so Brad went to the internal NASA launch control website to check on the tropical outlook. So far, so good. There was a tropical depression off of Puerto Rico, but that was over a thousand miles away. Currently they were calling for a good weather window.

Brad made a quick call to Meryl, an old friend at ISS's Ground Control. "Hey, Meryl, it's Brad over at CAPCOM, MMI. How are you guys holding up?"

"We keep doing our damnedest. How about you guys?"

"Same, I guess. I stuck a message on the capsule going to the ISS asking the crew to contact MMI if they had indeed been in communication with them as CAPCOM had theorized."

"Yeah, I heard," Meryl said. "Who knows? All we can do is hope for the best and stay tuned."

"I suppose. We all kinda feel like hamsters running furiously on that circular treadmill. Running like mad, but just not going anywhere."

"That's just how we feel, ol' buddy."

"You guys looking good for the launch on Thursday?" Brad said.

"Yeah, if the weather holds. I wish we could move the launch up another day. According to Launch Control, there's a tropical depression off Puerto Rico and it looks in favor of development."

"Yeah, I was wondering about that. A tropical storm can develop into a hurricane in twenty-four hours easily enough. How can they tell what the hell the weather is doing without any satellites?"

"Mostly radar, I guess. Of course, the NOAA weather planes fly if they think we need their hurricane hunters with all of their super sensitive equipment. The US Coast Guard has a base in Puerto Rico, too. I'm pretty sure that's too far away for any radio comms. Hurricane Maria took out most of the power grid in Puerto Rico, as well as their radar back in 2017. So, they have a fancy, state-of-the-art, super-powerful doppler radar system. I heard they went to solar power all over the island in a big way after the storm and the earth quakes of 2020. Maybe that kept them online. I really don't know. Anyway, we're off generators now, so that's a big relief. I can actually go home at night as long as I have a landline. I also carry a radio with me everywhere I go."

"Yeah, Meryl, I have one that goes home with me as well. When I do go home. We don't have a landline or food at my house. Thank God they feed us here!"

"You got that right, buddy."

"Okay," Brad said. "I just wanted to check with you on my message for MMI and the weather outlook for the launch. Let me know if the weather is going to throw a wrench into our plans. Good talking to you, say hi to Brenda for me."

"Ok, I will, and say hi to Susanne and the kids for me as well."

"You got it, bud. Let's hope the weather stays at bay for a clean launch on Thursday." Brad hung up the telephone and stared at it as he thought about the conversation. There's that word "hope" again. We are getting that a lot lately.

31. ISS - Day 13

Giulia opened her eyes to a blurry, distorted image. She couldn't see anything. She wiped her eyes with her sticky shirt sleeve. She was in her sleeping bag on the ISS. Something was wrong. She was covered in sweat. She could hear the guys stirring, so she crawled out of her sleeping bag and grabbed a towel. Then she looked around.

"Good morning, sunshine," Henry said. "Breakfast will be ready in a minute."

"Morning. What's the temperature in here?"

"You're on the hot side at the moment. I think the thermal insulation may be damaged on your side of the station, which is facing the sun right now."

"Oh, my God. It must be a hundred and twenty degrees in there."

"Yeah, we're about to enter Earth's shadow. It'll cool off soon enough."

Commander Amhurst piped in. "Yeah, we're not in a 'normal' controlled orbit of any kind. The station is slowly rolling around as it's orbiting. We have no control of yaw, roll, or pitch, so you're getting the sunny side right now. Welcome to day thirteen!"

Henry, Giulia, and Commander Amhurst floated up to the traditional breakfast nook to have some rehydrated fruit and cold coffee.

"Okay," Commander Amhurst began, "for our morning briefing, I think we should go over a couple of things. First thing this morning I did my normal rounds. The O_2 is back down to nineteen percent. We will open up the O_2 tanks right after breakfast. How that goes will pretty much determine if there's anything else we need to do. You guys have been great at minimizing water usage from the way it smells in here.

They all had a good chuckle from the commander's unusual joke.

Have you noticed anything or need to bring any mission-related issues to the table?

"Not me," Henry said. "You, Giulia?"

"Yeah, I need the air conditioning turned on in my sleeping quarters," she said, smiling.

After breakfast, Commander Amhurst made his way to the O_2 bottles and opened them one at a time. The bottles were noticeably quieter and losing pressure, and he shut them down after a few minutes. The commander then watched the atmospheric monitors for a good minute, tapping them on the monitor screen, then turned to look at his fellow astronauts. "Well, it's up to twenty percent now. It may climb a little more."

He didn't have to say anymore. They all knew they were getting close to having to make some difficult decisions. The air bottles didn't last as long as they had hoped. The CO_2 scrubbers never worked, either. They were from the early 2000s and had been sitting unused in storage for years. When the power was hooked up to them, they blew air okay, but no one could tell if they were working or not, and after a while, they quit blowing air altogether.

Commander Amhurst looked at his crew and exhaled. "I think we have three or four days, tops. Next time it goes down to nineteen percent will probably be the end of the bottles."

"I'm going to take that CO_2 scrubber apart, Henry said. "Wanna help, Giulia?"

"I will. What the heck."

Upon further analysis, they discovered that the carbon dioxide absorption canisters had been wrapped in a plastic film, which should've been removed prior to being inserted into the blowers. The plastic had wound around the fan and burned out the fan motor. They probably would have worked if the film had been removed.

Henry and Giulia looked at each other with downtrodden faces. They knew the scrubbers would've helped a lot. Henry thought to himself, this is the kind of thing that saved Apollo 13.

Henry disconnected the power and slowly unwound the plastic film that was wrapped around the fan. Then he reconnected the power and tried to coax the fan motor back to life, but it wasn't going to happen. "Well, I guess that's that."

Giulia agreed and went over to the atmospheric gauges. They still read 20% O_2. She looked at Henry. "Go fish?"

Henry's sad eyes said, What do you do when there's nothing you can do?

Somehow Giulia understood.

32. Mars - Sol 13

As Barbara Black continued researching her findings from the cave, Commander Carrigan was suiting up for another EVA.

Carl had already finished the scaffold frame and was out adding strengthening crossmembers, to add, as he put it, "rigidity" to the final construction. Commander Carrigan checked the fresh-water tank in the hold first thing that morning. It was almost empty. He estimated they had one or two more days before they were out.

Water restrictions had been in effect for two days. He guessed that they were down to less than one hundred gallons and the sensor meter or float showed nothing. This may sound like a lot of water, but twenty one people use approximately twenty one gallons per day on a normal day just drinking. Almost all of their food required water to be rehydrated. Without water they really cannot eat. The days on *Staten* were anything but normal. His crew works hard especially when on an EVA. Breathing air from a compressed air source dehydrates you faster than normal. Commander Carrigan had sent out an email to all crew members detailing the water restrictions, and he also placed it on the bulletin board as well as announced it in the crew meeting two days earlier.

Commander Carrigan hated making any announcements over the all stations PA because inevitably, depending on their shifts, some crew members were sleeping in their bunks. But running out of water was no joke. The crew was starting to smell a bit ripe, their olfactory systems would have to endure the insult until they had their water tanks refilled.

Carl had used stainless steel bolts and nuts to hold the scaffolding together every 30 feet, which allowed disassembly for moving it to the different starships. To make the project easier to manage, he welded the base structure at *Starship IV* and then moved the welding equipment and all of the supplies to *Starship VII*, where he finished welding and manufacturing the rest of the scaffolding. Fortunately, *Starship VII* wasn't far from *Starship IV*. Carl was kicking himself for not thinking about this before welding the first 20-foot frame together, but it didn't take long to move it the short distance in the lighter gravity.

Commander Carrigan asked Mary and Andrea to accompany him, and he asked Steven, who was running

Staten Command, to raise Carl on the radio and get him to bring one of the two Mars cars over to pick them up.

When they arrived at the bottom of *Staten's* crane, Haratu Suzuki was already waiting for them. Haratu ran them over to *Starship VII*, and as they approached, it was evident Carl had done his magic with the scaffold.

The commander walked around the huge structure and was amazed. Carl had found some flat square aluminum to use as feet. He had somehow wound the rope around the starship, tying it off to the scaffolding to make it secure, and then, instead of cutting the rope, he ran it farther up the scaffold, secured it again, and ran it around the ship until he ran out of rope. Then he used a different length of rope, so it didn't have to be cut. Commander Carrigan thought it smart, then wondered how the hell Carl got the rope around the starship, up the 50 meters to the top, and at each level where it was tied off. He didn't see Carl anywhere, so he assumed he must be up top.

After hailing Carl on his helmet mic, Carl confirmed that he was up top, so Commander Carrigan started climbing. Even with the reduced gravity of Mars, Commander Carrigan weighed around 150 pounds with his gear on, and after 60 feet, he started huffing and puffing. At 90 feet, he had to take a break. At 120 feet, he took another short break with only 30 feet to go. He could see Carl was just finishing tying off another length of rope which had been wrapped around the giant starship.

Commander Carrigan stopped to take in the view. Carl and his team of four astronauts had done the whole thing.

Don't ask me how, he thought.

Commander Carrigan was still trying to catch his breath. "How did you do this, Carl?"

"One piece at a time." smiling, and obviously proud of his work, Carl made a big performance out of it, as if he were revealing the scaffolding on The Price is Right.

Once Commander Carrigan pulled himself up the last three steps and landed on the platform, he was amazed. Carl had put the ladder at a comfortable angle, and every 30 feet where he tied off the scaffold, he had created a rough deck, which he wrapped the cargo netting around to act as a type of hand rail or safety netting. But it was apparent to the commander that the cargo netting was also adding tension to the structure, helping to stabilize it and keep it from wobbling.

"Wow, you must have used almost all of the rope."

"Pretty much. What's left will be used as the guide rope for the crane."

"Did you make any safety belts or tie-ins for you and your team up here?"

Carl turned around and showed Commander Carrigan the tie-in with a stainless steel snap link attached to his Mars suit. "Yeah, we have them. They're not to go up or down without tying in, and they swap them out with the guys below if they're going to change positions. We have a lot of these snap links from *Starship IV*."

"Excellent. You have done an incredible job. Carl." Commander Carrigan took a look around. "A hundred fifty feet in the air on Mars. Who would've thought we'd have to do this! I don't know how you did it, Carl, but it is amazing. And very sturdy. It's hardly moving."

"Yeah, well, it weighs a lot. I used two of the stringers on top of each other to give it a double thickness for extra strength on the first two levels, and of course I reinforced the frame with those crosspieces that look like giant x's. That adds to the strength. Plus, tying it off every thirty feet to the starship adds a lot of stability also."

"Well, once we get the water flowing, we'll throw a party in your honor."

"Don't do it for me. All I did was a bunch of welding. Do it for my team. After we got started, they understood everything I needed. I hardly had to tell them anything."

"Have you tried the hatch yet?"

"Was just about to do that, but I wanted to wait for you."

"Very considerate of you, Carl."

With that, Carl inserted a crowbar-like tool with two large teeth into the access and popped out the manual-release mechanism. In the manual release were two large holes, in which he inserted the crowbar apparatus and began to turn counterclockwise. After six turns, a hissing sounded as the hatch seam popped, and they swung it open as the lights inside *Starship VII* turned on.

Carl held the door for the commander, and they stepped into the artificially lit interior. It didn't have the equalization room just inside the door as *Staten* did. They were met with a flat deck large enough to hold five or six people. Just below them were aluminum shipping containers with heavy shackles for attaching the hook from the crane, and there was a walkway to the helm with a low railing. The helm had two chairs attached to the floor where the computer programmers would sit while installing the navigational and control computer attached to the varied instruments and monitors required.

Commander Carrigan examined the set-up as Carl crawled under the computer and turned on some lights underneath the helm. Commander Carrigan took a mental note of the layout and walked back to the platform to assess the crane. He found a remote control inside a control box— just where it belonged. He punched the power button. Nothing. He tried it again. Still nothing. Carl's sudden voice made him jump.

"Yeah, it's not going to work without booting the computer system up. Just as I thought."

Commander Carrigan nodded. "Yes, that does appear to be the case. No manual boot-up instructions hidden away or under the console either, I guess."

"Haven't seen any yet, but I will look around before tearing into that crane. I'm hoping I can run some wires to the motor with that remote control inline. That would almost be as good as firing up the computers." Carl paused. "Okay, maybe not that good. This is still going to be a lot of work. But impossible without the crane."

"Do you think it'll let you do that, Carl? I mean, do you think you can activate the crane?"

"I don't see why not. The computer system is designed to follow the programming they showed us in training. It's just a dumb computer. It doesn't know how to think outside the box, so to say. It doesn't have anything like self-defense mode or sentry mode or an automated lockdown. It already is locked down as far as it's concerned."

"That's great news, Carl. Do you know where the water hose is located?"

"I'll get the bin and location number from Steven. Let's hope they're not at the bottom. It could easily take two days or more to off-load this bad boy. If I can somehow get to them quicker, I will. I'm hoping to have the crane operating before the day is out today. I just have to get my tools up here." He looked down the 50-meter scaffolding."

Commander Carrigan gazed down the height of the scaffolding. "Well, whatever you do, be careful. Do you have any rope left?"

"I'll have to check. There's supposedly some more on this load, along with the third Mars car, and of course the water line."

"Do you need any more help?"

Carl looked back at the helm and the dark screen of the computer monitor, deep in thought. "If I can't find another two hundred feet of rope, I might."

"Let me know via the radio. We'll head over to *Starship II* and start stretching the water line out. That way when you get to the water line, it'll be a simple job of just connecting in the last one thousand feet. Let me know personally as soon as you're able to pull it out, please."

"Yeah, no problem. If I need more hands, I'll tell you as soon as I have it figured out."

"Actually, Carl, I can get you four more personnel suited up immediately. If you don't need them, you can cancel."

"Ok, Sounds good."

Commander Carrigan raised Steven on his helmet mic and gave the order for four of the five crew members of C shift to mobilize. They were to report to Carl as soon as they arrived at *Starship VII*.

He took Mary and Andrea to *Starship II* to start unbuttoning the water lines he had dropped off two days ago. He then helped transport C shift to help Carl, and then returned to *Starship II* to help finish stretching out the water lines the mile back to *Staten*. They had all of the bundles free, and the first 1000 feet was already hooked up to the water tank on *Starship II*. The first 1000 feet had been laid out for rapid deployment in a huge zigzag shape. Slowly driving toward *Staten*, they began stretching out the precious water lines, being careful not to let them kink. They stopped every couple hundred feet or so and made a pyramid out of the largest rocks they could move to mark the water hose line. This was about 25 stops, so it took a lot of time. Commander Carrigan made a mental note to dredge some small ditches to put the water line in. He thought, *After we get water running, I am going to ask Carl if he could make some flag poles or markers or something, so we don't accidently drive over the water line and put a hole in it.* He was imagining one of the Mars cars running the water line over creating a hole in the line and the water blowing out of the hose in the 1% atmosphere of Mars would probably create a geyser a hundred feet high as it turned into a cloud of humidity. *No, that we could do without,* he thought.

They made it most of the way back to *Staten* with the hose running short only about 1000 feet away. 1000 feet is like two city blocks! Frustrating. After making sure the bitter end was taped up, Commander Carrigan, Mary, and Andrea drove back to *Staten*, as the sun was about to set. Commander Carrigan asked Mary if she could take the Mars car back to *Starship VII* and return with the crews. Commander Carrigan confirmed this with Carl.

For the time being, *Staten* had a policy of no one out after dark. The temperature drops rapidly. They didn't need anyone getting lost (even though *Staten* was the tallest lighted object on the planet, not to mention the only lighted object). Still, if they could help it, Commander Carrigan didn't want anyone out in the dark and Mars is very dark at night. The only reason to be out after dark was to take in the incredible night sky. Which he agreed they all needed to do in the near future. It really was out of this world. And you could see the Earth most nights. What a feeling that created.

After equalizing, Commander Carrigan quickly made his way to the tiny lab Barbara Black jokingly called her home.

"How's it going?" Commander Carrigan said with a smile.

"Almost all of the diatoms are dead. Oh, well, it was to be expected, but I found DNA. They're using DNA, just like us! And I found a few different forms of carbon and other organics in the water. I know there's more to find in that cave.

You could easily find these same organisms somewhere on Earth! When can we go back? I have so much more to do and see, like the water source and the ceiling. I think it was a lava tube more than a cave and you know it keeps going deeper. When are we going?"

"Wow! Hold on a minute." Commander Carrigan couldn't believe her energy. She was like a little kid in a toy store for the first time.

"We went quite a way without much of a plan and no safety equipment. The next spelunking expedition will need to be planned out with safety backups," Commander Carrigan said. "You know we shouldn't have gone as far as we did without at least a safety line."

"I know, I know. We can do all of that on the next trip. So, when do we go?" She looked at Commander Carrigan as she clasped her hands together in a quiet applause.

"Soon," was all he could come up with. He wondered if Carl would find more rope in *Starship VII*. "We'll have to organize it, but first, we have to get water. If you didn't know, we're almost out of water now. I'm thrilled you found DNA. Have you been able to type match it to any earthborn DNA?"

"No, I haven't had the time to match it yet. At least the computer hasn't found any matches that I am aware of. I just started the search about thirty minutes ago. It's kinda complicated," she said, peering into her microscope.

"Further exploration of the cave is of course, a priority. Water is our top priority at the moment. If you organize the equipment, including about one thousand feet of safety line, I'll let you know when we can put it on the schedule. Probably in a few days, as soon as we get water." She smiled ear to ear, then went back to her microscope. Commander Carrigan left the little lab feeling like he had just had a conversation with a five-year-old.

On his way to the cargo hold to check the fresh-water levels again, he heard Carl and his crew returning from their work on *Starship VII*. Making his way toward the water tanks, Commander Carrigan entered the dungeon-like cargo hold, and the lights automatically came on. The fresh-water tank was registering "empty". Well, that will be the last time I check it until we get water coming back on board he thought. He wondered if they could last one more day, maybe if they reduce consumption to eight ounces per day? It was impossible to tell since the water level was now below the gauge. It was anybody's guess. "Boy, I sure hope Carl finds the water lines soon," he said out loud to himself.

Commander Carrigan sat down on the steps of the cargo hold and imagined crews having to carry water containers out to the water line to fill them up, then realized that the water would evaporate and boil off, even before they could get the caps back on the containers. He sat for a moment, trying to think of something else they could do to get water.

The water would have to go into a self-bleeding pressurized container, straight from the 5000-gallon tank at *Starship II*, or it could go into a low-pressure container that became pressurized as the water was added. Still, it was the same problem. They didn't have any containers like that, which would need to be specially manufactured for the job. The container would have to have an airtight seal between *Starship II*'s tank and the container, and the container had to be rigid, because of the low atmospheric pressure on Mars, it would blow up like an overfilled balloon. It might even explode.

Commander Carrigan now understood why the water line was braided stainless steel with heavy-duty connections. *That's also why it weighs so much,* he thought.

"I have to talk to Carl!" he said out loud to no one.

It was now dark outside, and some of the crew members were preparing dinner with what could be their last meal until they get water. The cook and kitchen positions were designated on the roster, but some of the crew members enjoyed making dinner, and they volunteered to do it.

Commander Carrigan went to talk to Andrea, who was finishing her daily report on the EVAs.

"Hi!" he said, startling her.

"Hi!" she responded a little too quickly.

"How you doin'?"

"Nothing new to report on the EVAs."

"Okay, good!" He liked the fact that she could read his mind. "Everything go as smoothly as expected?"

"Yeah, no issues." she smiled again.

"Okay, great. Thank you, Andrea."

Commander Carrigan asked Carl to speak to him in his office. The commander didn't have to ask him anything.

"We got the crane powered up, and we pulled the first container out," Carl said. "After guiding it to the ground with the rope, we opened it up. No water lines in this first one. Not too surprising. Steven said the water lines were about halfway into the load. There are fifty containers to each starship, and they're tightly stacked on top of each other.

There's no way to get to the middle of the stack without taking the containers off the top. Thank God *Starship IV* wasn't loaded using containers like these. It would've been a real mess."

"Do you think you'll be halfway into the load tomorrow?"

"Barring anything unforeseen, that's exactly what I'm hoping to achieve by the end of the day tomorrow."

"Ok, sounds good Carl. Thank you.

Commander Carrigan announced a full-crew meeting, knowing they were becoming entirely too frequent, but he really had no choice. The developments were life-threatening, and his crew deserved to be prepared.

"I'll keep this as short as possible," he said as soon as everyone, including Pedro with his crutches, sat into a chair.

Pedro's knee was in a brace, locked into position. He was still sleeping in the medical bed because it was easier to get in and out of compared to his bunk, but he seemed to be in good spirits.

"How's the knee, Pedro?"

"Pretty good. It still hurts if I move it, but I can tell its healing. Give me a couple of weeks, and I'll be back out there."

"Good to hear." Commander Carrigan said with a smile, then he glanced toward Tammy, who just smiled with a slight shake of her head. "The reason I've called us all together is rather serious. To get right to it, we're at the end of our fresh-water tank. As you know we've been surviving on the stores we brought from home and the fresh water that the hydrogen fuel cells were producing, as well as the ship's recycling systems. The hydrogen fuel cells need to be refreshed, but they don't produce enough water to be a significant source. Carl and many of you have been valiantly working on gaining access to the last bit of water line we need to finish the run from *Starship II* to *Staten*. Mary and Andrea and I have run five thousand feet to within maybe one thousand feet of *Staten*. I've been meaning to ask Carl if one of you guys can weld some markers for the water line, so we don't accidently run over it."

Two hands immediately shot up.

Commander Carrigan looked at Carl, who nodded in approval. "Great. Can I get you guys to take care of that in the morning?"

The two volunteers readily agreed.

"Thank you. The water line is listed in the manifest and inventory at halfway down the containers, inside *Starship*

VII. So far, we have one container out. By my calculations, we have twenty-four more, give or take, before we should have the one container with the badly needed water line. Carl, how long would you estimate it takes to hook up to the container you removed today and safely off-load it on the surface, then recall the crane back on board?"

Carl looked a little hesitant and thought for a few seconds as he stared at the ceiling. "Twenty to twenty-five minutes."

"Would you say that's a good rate to safely maintain?"

"Yeah, I think we could probably average twenty-five to thirty minutes per container. It takes a good ten minutes to pick one up and get it ready to guide it the hundred fifty feet down, and then another five minutes or less to recall the hook with the guide rope. So, yeah, no faster than twenty minutes, and some may take thirty."

"Okay, Carl says about two per hours. Whatever rate we maintain, I want to be sure we are staying safe in our work practices. We're told by the computer inventory spreadsheet that we must get to container number twenty-five. If our rate is two per hour, that means that we have a long, ten-hour day of solid work. We may get twenty more containers off-loaded, which means we'll be short four containers of getting to the badly needed water line, or about two more hours of work. If we stick to it for two more hours, we may get to container twenty-five, but we'll be out of daylight, and we'll still have to find the hose and hook it up to the water line, attach it to *Staten*, and turn it on at *Starship II* without running over the water line with the Mars car or steering into a crater or a ditch in the black of night. I just want everyone to see the picture here. We can start the day just prior to sunrise and continue to rotate crews every four hours. We'll be out of water sometime between now and by the time you wake up. We're really at the bottom of the barrel. I'm further reducing our water consumption to eight ounces of water per day, per person, though I don't think it's going to make any difference. Out of water means out. All of our foods must be rehydrated. Anything we drink must be rehydrated. Because sugar and caffeine both cause dehydration, they're both banned until we have water—and you know how much I love my coffee. I suggest we go with fruit rehydrated in our mouths for breakfast—if we have any water. I have to make some difficult decisions about tomorrow's workday, and I'm open to your suggestions."

Silence followed, then Wayne raised his hand. "Commander, how many people does Carl need to keep the unloading going at the rate of two containers per hour?"

"Carl?" the commander said.

"Including myself, seven. Two up top and four on the ground. With me acting as an observer."

"And how difficult would you say it is to learn each position?" Wayne asked. "How long would it take any one of us to learn the different jobs?"

Carl answered immediately. "The work isn't hard to learn. The positions up top have to push the container around to get it out the door. The person on the guideline has to be sure the container clears the scaffolding without getting tangled up, and the other three on the ground have to coordinate with the guy on the guideline, pick up the container, and move it to staging. I estimate each person will have to manually pick up about one hundred pounds using leftover aluminum framing from *Starship IV* as handles. They will go underneath the container that we'll place on some rocks so it's off the ground. It should take maybe thirty minutes to learn each position just by watching. The three guys on the ground, not on the guideline, are there only to pick the container up and move it to staging, but they cannot move it until the guy on the guideline allows me to get the hook back inside."

Wayne said, "Sunrise is at what time, Commander? About fifteen hundred Zulu?"

"Yes."

"So, if the first six of us start at fourteen hundred Zulu, we should be at *Starship VII* by fourteen thirty. As the sun is starting to rise, we can work until, let's say, eighteen hundred, when six more come to swap us out. Let's say we give them thirty minutes to learn the drill and we come back in at eighteen thirty. They can work until twenty two hundred when another six come to swap them out, and they get the thirty minutes of training. So, they're working from twenty-two hundred to O-two hundred Zulu, or whatever it takes to finish the job. Four straight hours per shift with no breaks. That leaves us with two extras, in case breaks need to be taken or someone has to be swapped out. With twelve hours of uninterrupted work, we should be bringing down the twenty-fifth container by the end of the day and you already have one on the ground." Wayne looked around, proud to have whipped that off the top of his head.

That's what Commander Carrigan was hoping to hear. Before the room broke into conversation, Commander

Carrigan spoke. "I think that just might get us water by the end of the day tomorrow. If the last crew sends Mars car one to *Starship II* to turn on the water once the water line is connected by the other crew in Mars car two, they may even make it back to *Staten* before sundown."

It was all in the math and shift rotation. Pedro announced that he would run *Staten* Command, which would free up one more person.

Barbara Black was heading to the galley, she was at the sink when she yelled out, "It's gone! It's gone! No more water!"

The room went silent.

"Okay, people," the commander said. "I suggest each one of us go to the ice maker and get one piece of ice. That will be our water for the night. Ashanti, can I get you for night watch, please? And we can all enjoy one piece of ice in the morning, and one piece after our four-hour shift, if it lasts. With any luck, we should have plenty of water tomorrow night."

33. Johnson Space Center - Day 14

In Brad's mind, the news couldn't have been much worse. The tropical storm off of Puerto Rico had turned into a full-blown hurricane and was rounding the northern tip of Andros Island.

The name of the hurricane was Brad, and it was looking to be a category two by the following day, with winds in excess of one hundred miles per hour. Currently about one hundred miles wide, it was causing havoc from Cuba to Miami to Andros, and soon Bimini, with a forward speed of sixteen miles per hour. It was projected to head due north, up the east coast of Florida, with Ft. Lauderdale in its sights. If the forward speed of progression continued, it would be in Cape Canaveral the following day.

Brad picked up the phone. "Hey, Meryl. It's Brad. I guess you know why I'm calling."

"I could take a wild-ass guess," Meryl said with his usual sense of humor.

"You'd probably be right," Brad said. "You ever heard of a hurricane called Brad before?"

"Can't say that I have, buddy. But then again, they always retire the ones that do any damage. You may be interested to know that due to the hurricane, they've decided to move the launch up to today. It was fueled up yesterday and early this morning. We have the launch scheduled in T-minus three hours and counting, scheduled for fifteen hundred Zulu. That means the crew could be back on Earth by 6:00 p.m. Your wish may be coming true my friend. With any luck, it will go today. It has to go today, or it probably won't go for a week. Cape Canaveral locks everything up tight for hurricanes. I'm actually surprised they're a go for the launch. Have you spoken with our boss lately?"

"Only to keep him apprised of the situation, as usual. I have spoken with General Latimore a few times, but only to coordinate the message for MMI and to get his consent, this being an armed forces mission and all."

"Okay, I was curious," Meryl said. "I guess they're far enough along to get it off today. I hope those guys are still on the station."

"Well, their Soyuz capsule is reportedly still there. So, we are hoping they are as well. Have you heard anything about the Russians? I heard they left in their Soyuz capsule."

"Yeah. No comment from our Russian friends. With all of the chaos in the world, I'm sure they're hoping that it'll be missed in the thirty-minute attention span of today's news cycle. There is plenty to report on these days."

"Wow, no comment, and three cosmonauts dead. What a country! What must their families think? These guys should be heroes trying to push their country forward in the exploration of space, not swept under a rug like an embarrassment."

"Yeah, I know. I wonder if they'll even get a funeral."

"Okay, Meryl, thanks for the update. Let's give our guys a chance."

"You got that right, bud. They deserve whatever help we can give them."

Brad stared at the handset as he set the phone down. What must it do to the morale of the armed forces and cosmonauts to know that their comrades died, and that Mother Russia swept it under the rug? The next thing they know, they'll be saying the cosmonauts didn't have permission to leave the ISS and were told to stay, so their deaths are by their own hand.

Brad turned to his computer and pulled up the NASA in-house net, then scrolled to the launch window. T-minus two hours and thirty-eight minutes. It looked like a beautiful day in Cape Canaveral. Brad decided to grab a sandwich and soda and watch the last hour of the launch as he went over his notes.

Had it really been two weeks since he spoke with the MMI crew? Brad had commissioned himself and all MMI staff to start imagining what the *Starship X* crew was going through and what he and his staff would do if they were in the same predicament. "I want any imaginative stuff you guys come up with, I don't care how far-fetched and crazy it may seem. If it 'could' work, if it has the smallest of possibility of working, I want to see it. Make drawings or designs on your CAD/CAM or drawing software or by hand. I don't care. How are they going to access the one hundred and fifty foot hatches on their supply ships? That is what we need to know so we can help them. So far, they had projected that the crew on Mars would've probably off-loaded *Starship IV* by hand using *Starship X*'s 20-foot ladder. That would've given them one thousand feet more of water line, but not enough. They should be running out of water any day now so that gives them three days to get the water from *Starship II* back to *Starship X.*

They had no idea what *Starship X* would do for water, except maybe refurbish the hydrogen fuel cells, which would deliver some water, but only the bare minimum—not enough to keep twenty-one crew members alive for long. Brads team had considered using the ropes and lines from *Starship X* and *IV* to somehow lasso the huge starships, maybe using the hydrogen in a large plastic bag as a balloon to float a line over the top. But that seemed impossible to control. They considered stacking the Martian rocks alongside the starships using the Mars cars and trailers, but the sheer number of rocks was confounding.

His team estimated they'd need around six thousand rocks to reach the hundred-fifty-foot-high hatch, which they figured would take six months to assemble. At this point, he and his team decided that the twenty-one astronauts on *Starship X* were living on the edge of dehydration. Man cannot survive for more than three days without water. If they were dehydrated, they certainly couldn't carry a bunch of rocks. Just the simple act of breathing would dehydrate them, and all of the food on *Starship X* was dehydrated, which would dehydrate them even faster—if they ate.

They considered the possibility of tipping a starship over to gain access, using a jack or lever of some kind, but that was dismissed. It didn't seem possible. They considered that the crew might use shovels to dig under the huge platform like landing pads or feet of the rockets, thereby tipping them over by undermining them. Many of the starships had containers inside them, stacked like poker chips. If the rocket was knocked over, the containers would smash together and be almost impossible to remove. This action would jeopardize many aspects of the mission, but it would give them the precious water lines, if they were not damaged in the process. Unless the starship should roll over onto the access doors. Ugh!

So, this avenue was at the moment thought to be the highest probable action. All planning revolved around what seemed to be the most logical idea. Trying to figure out which starship the astronauts would try to knock over was a toss of the dice. There were three different starships to consider. They decided to draw up different outcomes for undermining each of the three starships that had water lines. The outcomes were all pretty much the same in the end: some damaged equipment, but they could still access the water lines, which is life-saving at the moment.

Different strategies were drawn up on how to move forward with the mission depending on what equipment or supplies were damaged. This gave Brad the chills. They may damage one of the nuclear Kilopower units, and they were sure to damage one of the three Mars cars. There had to be a better way. Brad hoped they found a better way into those starships, but he wasn't sure how they'd do it.

At T-minus twenty seconds, the countdown stopped. After the seventeen seconds it took for the countdown to restart, Brad could breathe again. Finally, the gigantic ship, with its super-reinforced payload, launched, belching flames as it slowly gained altitude. God Bless, and Godspeed Brad thought.

The flames lit up the inside of a foretelling cloud and disappeared downrange, rolling to its correct azimuth for orbit around the Earth.

34. ISS - Day 14

The inner hatch was closed and locked behind Commander Amhurst. Giulia and Henry heard the slow hissing of their precious air as it was equalizing in the airlock to the vacuum of space.

"There he goes," Henry said to Giulia, who was looking a little strained.

Who could blame her? Five days in a space suit, locked in a capsule the size of a Volkswagen Beetle with two other guys. In this tight confinement even just a leg cramp is a major event. Then almost ten days in a cylinder with the same two guys, never knowing when the air was going to run out or when the next round of debris would kill her.

Henry had to give it to her for being such a trooper. "If I weren't already married, I would be tempted to ask you to marry me. What the hell, I think we're way beyond our precious policy and procedures and protocols by now."

Giulia smiled inwardly and outwardly. They had become more than co-workers. They were very good friends. She truly loved these guys, and she knew they loved her and would give their lives to save hers. She would've done the same. As difficult as it may sound, it was true. Living that close together, they might as well be married—just no sex. At least they weren't having sex. There were rumors about some previous astronauts. Living that close together, each person knew the physiological intimacies of the others around them. In zero G, the stresses on their bodies were terrible. They knew if someone had diarrhea or gas. They knew if they were hungry. Everyone had body odor—some just smelled worse than others.

She laughed. "Can I marry both of you?"

They floated over to the windows and turned up the radio to listen to Commander Amhurst, who was just exiting the outer hatch.

He was giving the play-by-play, so the others could hear what he was seeing. "Exiting the hatch now and going up top to get a better view. On top of Unity now, looking at the damage to her. I'm surprised we have any pressure inside at all. Unity is pretty banged up. A lot of the impact protective jacket is floating around in loose patches here and there. Oh, my God. I don't know where we're getting power from because the solar arrays are all blown apart. Yeah, so are the heat radiators. That is amazing. The only solar arrays still here are free-floating and attached only by the power cables. That's probably the occasional scraping sound we're hearing. I can see the Harmony node now. Yeah, numerous obvious strikes. It's not looking good, guys. Okay, I'm making my way over to the Zarya and the Zvezda nodes now. Coming over the top of Unity. The Russian Soyuz is gone, but we already knew that. Oh, I can see both the Zvezda and the Zarya. I can see the Zvezda—or what used to be the Zvezda—best from here. I'm afraid there's not much left of it. It's completely open to space on one side. I guess there's no reason to go any farther. Oh, I can see the Zarya a little better now, too. It looks somewhat intact from this side, but I don't need to go any farther after seeing the Zvezda. I guess I'm coming back. No reason to waste more oxygen. I guess we have our answer. I'm back on top of Unity, heading to the hatch. Opening the hatch now. Heading back inside now. Closing the hatch. Okay, I'm in. Equalizing now."

Giulia and Henry could hear the hissing of the air as the airlock slowly re-pressurized from the onboard tanks. Henry floated over to the atmospheric gauges. No wonder he was feeling a little foggy today; the O_2 percentage was down to 15. They were waiting until after the space-walk was finished to add whatever oxygen remained in the tanks to their atmosphere. He had seen Commander Amhurst check the levels before suiting up, but he didn't say anything. Now

Henry knew why. Humans need a minimum of 19 percent oxygen for normal activities and 16 percent, minimum, to stay alive. Below 16 percent, the human brain and the cells of the body begin to go into hypoxia. Confusion begins, followed by rapid heart rate and rapid breathing as the body burns oxygen in a vain attempt to gain more oxygen and reduce the carbon dioxide build up. The lips and mucous membranes turn blue, as do the nail beds.

"How are you feeling, Giulia?"

"Besides smelling like I haven't taken a shower in two weeks, I'm okay," she said with a slight giggle.

"No, I mean mentally," Henry said.

"Maybe just a little foggier than normal."

Just then, the hissing noise stopped, and they opened the hatch to let Commander Amhurst back in. The first thing he did after removing his helmet was look at the atmospheric gauges. He was shocked to see the O_2 level at 14 percent. Commander Amhurst immediately went to the air tanks and opened them up. For about ten seconds, the sound of the O_2 could be heard as a slight hissing noise, and then . . . nothing.

He floated over to the gauges again. "Sixteen percent O_2," he said through a labored breath. "Well, that's the last of the oxygen. I suggest we hydrate. We have to move pretty soon here. How's everyone feeling?"

"Maybe a little more foggy than usual," Henry said.

As Commander Amhurst began to remove his space suit, Giulia floated over to give him a hand.

"As soon as the O_2 falls below sixteen percent again, that's it. No one can survive for any length of time at that level. We will have to suit up and load up in the Soyuz. I'm almost wondering if I should even bother to remove my suit. It may be only a matter of minutes. Boy, you guys should see the attitude of the Space Station. We are all over, end over end with a small roll to port. Fortunately, the motion is slow, but we will need to push away quickly in the Soyuz, so the station doesn't hit us in the butt as we leave." The commander began removing his suit anyway.

Henry checked the atmospheric gauges again. "Yeah, it's flickering between sixteen and fifteen percent right now."

Commander Amhurst sat down. "Okay. I guess that's it, he said as he puffed, struggling to catch his breath. We might as well suit up and get ready to leave. I'm going to radio *Starship X* to let them know our situation before we go." He was growing agitated and was huffing and puffing more than usual.

Henry changed the radio frequency to reach Steven on *Starship X.*

"Go ahead and raise them," Commander Amhurst said. "I'm going to sit here for a few minutes to catch my breath."

Commander Amhurst was looking pale. Giulia took his hand and pinched his fingernail between her forefinger and thumb.

"I'm fine," he said.

She inspected his face and pulled his lower lip out, noticing its dark purple color. "You're cyanotic," she said. "All of the strenuous activity outside combined with suiting up and equalizing. You need oxygen."

"I'll be fine. Just let me catch my breath." He puffed as he wedged himself into a corner, still in his space suit trousers.

"You may not be able to catch your breath in this atmosphere."

"Just give me ten minutes. If I can't catch my breath, I'll put my suit back on."

Henry picked up the radio mic and tried to raise *Starship X.* "*Starship X, Starship X*, this is ISS. How do you copy?" He grabbed a pencil and wrote down 17:05 Zulu, knowing they should hear back in about twelve minutes, assuming they were still within range of the ISS's weak radio transmissions.

They floated around, keeping an eye on Commander Amhurst. Nothing had to be said. They knew they were in a dire situation, but they didn't know the extent to Commander Amhurst's hypoxia. Giulia kept checking his nail beds, but nothing was changing.

He lay down, trying to relax as best he could. At 17:17 Zulu, the radio burst to life. "ISS, ISS, this is *Starship X*. We copy you two by two. Good to hear from you guys. How are you holding out? This is Pedro Lopez, flight engineer. I know you guys usually talk to Steven. He's out on maneuvers. I'm monitoring comms today. Over."

"Hello, Pedro, this is Henry Wilson on the ISS. We are about to evacuate the ISS in the Soyuz and wanted someone to know what we're doing. The oxygen stores are completely exhausted. Commander Amhurst did a space-walk today to evaluate the remainder of the ISS. We've concluded that the station is no longer habitable, and we intend to board the Soyuz for an attempt at reentry. Copy?"

Once again, Henry wrote down the time. It was 17:20.

Commander Amhurst was still losing color, slowly turning pail.

"Giulia, is it okay to sleep in hypoxia?" Henry asked.

"I am not sleeping," said a lethargic Commander Amhurst. "I'm just resting. You're doing a great job with the telemetry, Henry. Giulia, why don't you suit up? Henry can, too, as soon as his transmission is finalized."

Giulia went over to the atmospheric sensors again. "Reading O_2 level at fourteen percent. And the atmospheric psi is at twelve. Is the hatch sealed correctly? It can't be, unless . . ." She floated to the hatch and put her ear next to it, then she began moving to the repairs they'd done the first day they reentered the Unity node. One of them was making a small hissing noise. She looked more closely and saw that the edges of the duct tape they'd put over the repair job looked a little loose. She checked the edge of one of the repair jobs, and the duct tape pulled right off. The adhesive used on the tape wasn't holding. It was oxidizing and losing its stickiness, it also looked wet.

She immediately put more tape over the repair job, but it wouldn't stick well. As she was checking the other repair sights, she found two more were leaking, so she taped them over again. "I see," she said. "Our breath is causing condensation to form on the walls of the station. It's underneath the tape and causing it to lose grip. I'm not sure why I'm fixing these repairs if we're going to leave."

The radio chirped at 17:36. "ISS, ISS. This is *Starship X*. We copy you on your intention to evacuate the ISS in the Soyuz. We will enter this communication into our ship's log, as usual. Do you have a specific time and date of departure? I have relayed this information to Steven, who is out on the Martian surface. He asked me to relay a personal message, but first from Starship X. We cannot possibly know what you've been through and what you may face. We wish you the best of luck and Godspeed. Steven's message is '*Aut viam inveniam aut faciam.*' He said you would know what it means. Take care. You are in our prayers. *Starship X*, over."

Henry picked up the mic. "*Starship X, Starship X.* ISS, we copy you. Our intention is to depart immediately upon conclusion of this transmission. Best of luck with your mission as well, and we thank you. ISS, over." With the radio transmission concluded, Henry hung the mic back in its clip and turned to check on the commander. "We need to get him in his space suit and in some better oxygen."

Commander Amhurst could barely stand while Henry and Giulia got him into the rest of his suit. After putting his helmet on and making sure he had air flowing, they opened the hatch and helped him climb into his seat in the Soyuz. As they climbed into their space suits, they noticed the commander's health improving, and he was beginning the startup sequences for the Soyuz capsule. When they were ready to board, Giulia floated back over to the atmospheric gauges. The O_2 was at 13 percent, and the atmospheric pressure was at 11 psi.

"Still losing oxygen and air pressure," she said as she put her helmet on and locked it into place.

Henry looked into the Soyuz and asked Commander Amhurst if he wanted the middle seat as usual.

"Hell, yeah," the commander said, surprising them. "I wouldn't miss this for the world." He slowly crawled back out, allowing Henry to take his usual port side window seat."

Henry nudged Giulia. "I guess he's feeling better."

She snickered. "I guess so."

Henry crawled in, with Commander Amhurst and Giulia filing in behind. Will the low charged batteries or better yet discharged batteries launch the parachutes? They did not know, and that thought had not escaped them. What could they do about it? Nothing. So, it wasn't discussed. They had to go before the CO_2 scrubbers on their space suits were saturated, and they had to be close to that by now. They were out of options and the only options they had remaining were not good ones.

Giulia shut the hatch on the Soyuz and locked it down. "*Aut viam inveniam aut faciam!*" Commander Amhurst said aloud as he reached for the release.

"What was that?' Giulia asked.

"It means—"

"No, not that! I heard something on the radio in the station. Henry did you turn the radio in the station off?"

"No, I guess I didn't not that it mattered much."

"Maybe it did matter."

Hang on a minute." The Soyuz went silent as they heard a computerized voice.

"I heard it again!" she shouted.

The men were now silent. Ten seconds clicked by, then, "ISS, ISS, this is the unmanned rescue capsule Crew Thirty-Two. Prepare for automated docking and subsequent evacuation."

It was a computer-generated voice. That wasn't *Starship X*, so what was it?

Henry peered out his window at every possible angle.

Commander Amhurst commanded, "Open the hatch, Giulia!"

She was already doing just that. Giulia threw the hatch open and pulled herself out, then immediately went to every window as the guys were pulling themselves out. "I don't see anything!" she yelled.

They heard a quiet hissing, almost a popping noise, in rapid sequence.

"Those are cold gas thrusters!" Henry yelled.

The entire station moved slightly, and immediately after, they heard and felt something bump into the starboard hatch. They floated over to the hatch as it became dark, then black.

The radio chirped. "Crew Thirty-Two, good docking. Hatch secure."

Commander Amhurst opened the hatch to the seal of a beat-up but secure hatch on an obviously heavily reinforced crew capsule. It was the most beautiful capsule he had ever seen.

Giulia yelled, "I can't believe it! I can't believe it! I can't believe it!"

Henry laughed out loud, and Commander Amhurst joined in, saying, "I think I just died and went to Heaven."

It was so brightly lit. Their eyes had been in the artificial light they'd jury-rigged on the Space Station for weeks, with just the strange angles of the sun occasionally coming through the windows. In the seats were three envelopes with instructions. He passed them back to Henry and Giulia as he opened one. They were instructions to be relayed to *Starship X*.

"They must be important!" Commander Amhurst yelled from inside the new crew capsule.

"How did this thing make it up here?" asked Giulia.

Commander Amhurst said, "If it made it here, odds are it'll make it back. Yes!" He was grinning ear to ear and shaking his fists looking up.

Henry was back on the radio before Commander Amhurst and Giulia were out of the new crew capsule. "*Starship X, Starship X*, this is ISS. *Starship X, Starship X*, this is the ISS. Stand by for instructions from CAPCOM." Henry then read the instructions into the mic. When he finished, he repeated the instructions once again, then said, "Please confirm instructions."

Giulia and Commander Amhurst came back out of the new crew capsule as Henry declared, "Now we wait." The pencil was tiny in his bulky space suit glove as he tried to write down "18:19," the time of transmission, on a clear area of the ISS wall. He circled it to make it stand out.

Commander Amhurst let out a big breath. "I would have to say that our odds of making it home in one piece have just been improved significantly. I want to get a better look at that capsule from the outside."

There was only one window from which it could be seen, it didn't provide a very good view. Still, the view they had gave them a glimpse of the same kind of blanket the ISS was wrapped in, though in some areas it looked different. The crew capsules were usually white, and this one appeared to be black underneath the Kevlar covering blanket. It also looked bulked up. They could barely see the cold gas thrusters under all of the padding.

At 18:31, the radio squawked back to life. Henry grabbed his instructions, so he could follow them as an excited Pedro read them back to him from 35 million miles away.

"ISS, ISS, this is *Starship X*," Pedro said. "Repeat, ISS, ISS, this is *Starship X*. The instructions you transmitted previously, we copy as follows." He then read them word for word back to the crew on the ISS. When he was done, he added, I don't know if this is going to work, but if it does, it'll totally change our current state of conditions. We are out of water only because we cannot get the starships to turn on. If this works, it'll be like a miracle. I don't know how you came by this information, but it means that somehow, you've been in communication with CAPCOM or MMI. Please return form of communication. We are still in a radio blackout and can only communicate with you. Thank you. This is *Starship X*, over. Copy?"

"Commander, do you want to follow up?" Henry said.

"Hell no. You're doing a great job. Carry on, young man. I think I'll have me a seat," he said as he went back into the crew capsule and sat down in the commander's chair.

Giulia thought, I wonder if his CO_2 scrubbers are working correctly?

Henry picked up the mic. "*Starship X, Starship X*, this is the ISS. Your instruction copy is correct to the word. This information arrived at our location from CAPCOM MMI via a heavily reinforced crew capsule. It appears to have taken some hits, but it is intact, and it will be our ride home. Again, radio communication with ISS CAPCOM or Ground Control is still not intact. Again, this information came to us in written form on a heavily reinforced automated crew capsule. It's obvious to us they went to great lengths to get this information to us, so we could transmit it to you. It is our privilege to provide this information to you. *"Aut viam inveniam aut faciam!"* We will be evacuating the ISS upon completion of this transmission. May God be with you. Over. This is the ISS, signing off." Henry let go of the mic. "Let's get out of here."

Commander Amhurst laughed. "Damn good idea!"

"I couldn't have said it better myself!" Giulia said.

Inside the crew capsule, it felt to Henry and Giulia like they were about to take a ride in a luxury sedan.

The automated computer had a question they had to answer before they could do anything. It said, "Did you transmit instructions to *Starship X*?"

They pushed the pressure-sensitive screen where it said "Yes" and felt the capsule release from the ISS. But something was wrong. Something was beeping at them.

Then Giulia noticed something on the control monitor. "Put on your seat belt, Commander Amhurst!"

After Commander Amhurst put on his belts, the beeping stopped.

Giulia sighed. "Okay, I'll marry both of you, and don't give me that shit that you're already married."

As the crew capsule began reentered Earth's atmosphere, they heard different bangs, booms, scrapings, pops, and other mysterious noises. Some sounded as though they might destroy the capsule due to the severity of the hit. The crew capsule shook, vibrated, and even flipped end over end. For a while, they didn't know if it was going to recover.

Can it get the heat shield side down with all of this interference? Commander Amhurst thought.

Eventually, the noises faded away and they felt the heating of the atmosphere upon re-entry. The sound of air or wind building up and screaming passed the capsule sounded worse than a hurricane. They didn't know where they would land on Earth, but they didn't much care. They all felt they had a high probability of making it through the debris now, and when they felt the heating of the atmosphere, their hopes began to swell. When the humongous parachutes deployed and swung them back and forth, they could not stop screaming with joy. They had made it when only minutes ago, they were sure they would not. As the rescue capsule splashed down in a beautiful blue calm Pacific.

"Aut viam inveniam aut faciam". I will either find a way or make one.

35. Johnson Space Center -

Day 14

Brad's phone rang and caused him to jump. It hadn't rung in such a long time, he'd almost forgotten about it.

It was Meryl. "We have radar confirmation! The crew capsule is making reentry. Get on the radar website, now!" Then he hung up.

Brad didn't have to pull it up. Someone at MMI Ground Control had put it on one of the seventy-inch screens overhead. There was just no audio. The capsule looked beautiful as it streaked through the sky. The first drogue chute opened, leading to the opening of the oversized parachutes, then the capsule slowly drifted down to a calm Pacific Ocean.

After the inflatable boats launched, a helicopter circled overhead as the crew capsule auto-inflated an oversized collar to keep the heavy capsule afloat. A ragged crew, still in their full space suits, emerged and fell into the inflatable boats, where they were whisked off to the aircraft carrier.

Were they able to transmit the data to *Starship X*? Brad guessed that the low-quality images he was watching were probably being transmitted via the aircraft carrier to an

AWACS plane overhead. The AWACS plane was probably retransmitting via line of sight to Hawaii, then converted to a secure telephone line running 2,500 miles to the West Coast of the US, then converted to secure internal web feed.

It's no surprise that it's a little fuzzy, Brad thought. We had no idea how dependent we had become on satellites before this.

The carrier appeared to be a huge catamaran vessel, one he had never seen before. The inflatables ran between the two enormous hulls, the crew was helped onto a large ramp, and then onto a dry ship deck. For a few minutes, the screen only showed reruns of the splashdown. It then cut to a presentation ceremony where an exhausted crew was seated and in brand new NASA flight suits.

There was no press to welcome them home. They waved to the applauding crew of the carrier, however, due to the extended zero-gravity exposure, the entire ISS crew could barely stand on their own and were seated immediately once they were helped to the stage. It was obvious they were thrilled to be back on Earth's surface.

Brad was glad to see the crew and knew how lucky they were to even dock with the reinforced crew capsule, not to mention to return in one piece. But did they send the instructions? He was pretty sure they would have since there was no way to board the capsule without seeing the all-important message. But were they in contact with the crew of *Starship X* on Mars? No one asked the question. Of course, they didn't. No one knew about the mission to rescue the crew from the ISS or any message to the crew on *Starship X*.

Brad immediately picked up the phone and dialed General Latimore. His secretary said she would take his message and relay it to the general as soon as he was available. There was no way Brad could make a phone call to the carrier, and he was sure that his boss Gerald Adams was also watching this transmission and would also want this question answered. So, would General Latimore, wherever he was.

I wonder if he's on that aircraft carrier. Brad thought.

Brad changed the screen on his computer to the coverage of the landing, so he could get a closer look at the cheering crowd of Navy personnel. There was no audio—only video feed. He knew he would hear sooner or later, but the anxiety was going to kill him. As the crew began shaking hands with the different officers, he spotted General Latimore in the congratulatory line. The general leaned forward while he was shaking Commander Amhurst's hand and said something to him. Commander Amhurst nodded his head and pointed at Henry Wilson. When Henry shook General Latimore's hand, he nodded vigorously, and the general slapped him on the shoulder in a congratulatory gesture. They were both smiling. Was that the confirmation?

Brad was pretty sure that what he had just seen was good news. At least they weren't shaking their heads and looking down or upset. Brad considered how long it would take the general to fly back to the States. Probably eight hours or so. It was 22:05 Zulu, which meant 4:05 p.m. in Houston. The general wouldn't be back until almost midnight if he didn't stay too long for the ceremonies, which probably wouldn't last long anyway; the crew needed rest. He knew they would be going straight to sick bay.

Maybe the general will get his messages before going home for the night, which means I may not hear until tomorrow. Brad thought. Well, looks like I'm not going home or going to get any sleep tonight.

36. Mars - Sol 14

As soon as Pedro finished with the transmission to the ISS, he waited the twelve minutes for the confirmation return. Sure enough, at 18:45, *Staten's* radio came back to life.

He was thankful for the new computerized radios that prevented him from missing incoming transmissions. The radio was on and tuned to the right frequency, but Pedro was paying attention to the activities on Mars. Fortunately, the new radio lights blinked on and off as each new transmission came in, so Pedro didn't miss anything.

"*Starship X, Starship X*, this is the ISS. Your instructional copy is correct to the word." *Staten's* telemetry computer was recording every word and so was Pedro. As soon as he received confirmation back from the ISS, he was on radio Alpha to Commander Carrigan. "*Starship X* Command to Commander Carrigan."

"*Starship X* Command, this is Commander Carrigan."

"Commander Carrigan, switch to Bravo frequency."

"Commander Carrigan switching to Bravo frequency. What's up, Pedro?"

"Commander, you are not going to believe this, but the ISS just sent us the instructions from Mars Mission I

CAPCOM via written communication on how to fire up all of the starships."

Commander Carrigan was so surprised to hear this information, he forgot all communication protocol. "I'll be right there!"

Ten minutes later, Commander Carrigan was leaving the equalization room, heading straight for the helm, where Pedro was working at the telemetry station and monitoring *Starship VII* Command.

"I am not sure if I heard you correctly Pedro." the commander said.

"Yes, sir, they're right here, if you can read my writing. They are also recorded, of course."

Commander Carrigan read the instructions. "You have to be kidding me. It is that easy?"

"Sounds pretty simple to me." Pedro had to admit.

Commander Carrigan rubbed his forehead in thought. "We're about two hours ahead of schedule, which is better than we could've hoped. I'm not sure what would happen if we turn *Starship VII* on while we're in the middle of unloading it. Besides, Carl has disabled the crane from the automated computer system. I think we should continue manually off-loading the rest of *Starship VII*. But to get the other starships to autonomously off-load as they were intended is welcome news indeed. Is there any ice left?"

"No, we're completely out of water now."

"And the crew is out there working hard, too. Well, if all goes well, we should have the tanks filling before the sun sets," said the commander.

Commander Carrigan went back to suit up again. He needed to talk to Carl. After equalizing to the pressure of Mars, Commander Carrigan made a mental note to recheck the O_2 levels in *Staten*. They have been stealing away the kilopower unit to weld scaffold. On his drive over to *Starship VII*, Commander Carrigan realized that with the equalizing and water-focused activities taking place on the Martian surface, it must have taken a toll on *Staten's* O_2 levels. If it weren't for the MOXIE working so well, they would've been racing to attach the O_2 lines to one of the back-up tanks on *Starship I* or *II*.

Boy, we're still hanging by a precarious thread, Commander Carrigan thought. If any one thing stops functioning correctly, or if we don't get the water lines out today, we will be hurting . . . or worse. Just one day without water can sap a crew's strength and morale. Two days any meaningful work is almost impossible and on the third day your dead.

Upon arrival, Commander Carrigan looked around for Carl, who was, of course, all the way up top. He waited until the crane off-loaded another container, then headed up the 150-foot scaffolding. He did better this time, only having to stop once on the climb up.

I need to keep doing this, he thought. *This will keep me in shape.*

Commander Carrigan ducked inside as they were hooking up another container and getting ready to push it out the hatch. Wayne was operating the crane, and Adnan Ashari was manhandling the big snap hook set in place of the automated clinching teeth. Carl had explained earlier that it was faster doing it this way. It took more personnel to operate it, but it was about two minutes faster per container.

Carl was up in the helm, well out of the way of the work he was monitoring.

"How's it going?" Commander Carrigan said.

"I couldn't do it any better if I was doing it myself. Looks like we have somewhere between one and two hours before we get to the container that has the supposed water hose in it. We're averaging three containers per hour. Maybe one hour left, assuming the packing list is correct."

"That's great, Carl. We have more people out here than we need. Why don't we have a few of them open the containers and verify their contents against the packing list? That way, if it is accurate, we'll have more confidence in locating the water lines."

"Sounds good," Carl responded.

"I would like to talk to you on the surface anyway. I have a surprise for you," announced the commander.

When Commander Carrigan was safely on the surface, he hailed Pedro on Alpha frequency. "Starship Command, show Carl and me switching to Bravo frequency for a few minutes."

Pedro repeated, "Roger, Commander Carrigan, and ES White switching to Bravo."

Once they both changed frequencies, they dutifully signed in.

"Commander Carrigan on Bravo."

"Carl White on Bravo."

Pedro responded, "Carl White and Commander Carrigan on Bravo."

As they walked toward the two Mars cars parked near *Starship VII*, Commander Carrigan said, "Carl, suppose we knew how to activate the starships to go into autonomous mode. Which one do you think we'd want to start up first?"

Carl's eyes lit up. "Excuse me? You wouldn't be asking me this unless you knew something."

"You're too smart for me, Carl, and you're absolutely right."

Carl's eyes were the size of dinner plates. "You're kidding me, right? he smiled"

"This afternoon, we received a call from the ISS. Apparently, they had some reinforced rescue capsule arrive at the ISS that had a written message for us. The crew on the ISS relayed the message just prior to their departure. The message explained that if we connect the computer frame to ground, it should reboot itself. Can you believe that? All of this work, and all we needed was some jumper cables."

"You have got to be kidding me! I almost did that the first day you and I were on *Starship VII* together, but I didn't because I thought . . . well, I thought I shouldn't because I didn't know what I was doing. All of this sensitive computer shit. Oh, my God! So, why are we here? We should've tried it on *Starship VII* while we were there."

"I thought of that, but we're almost halfway done unloading it, and the scaffolding is in the way of where the crane will be unloading the supplies. Not to mention that you have disabled the crane," Commander Carrigan said.

"Hmmm, I see. I guess that makes sense. Besides, who knows what it would do now?"

"Exactly," said Commander Carrigan as he sat down in one of the Mars car seats. "Oh, this feels good. As I see it, we need to off-load the rest of the containers on *Starship VII* manually, just as we have been, once we obtain the rest of the water line we need. We can perhaps slow the rate of the work down. We don't want to stress the crew out any more than they are. And I have to tell you Carl, you and your crews are doing a phenomenal job out here. Once we refill *Staten's* tanks, we'll have some breathing room. But that's not why we're having this little chat." Commander Carrigan paused.

Carl's face clouded over as thoughts and ideas raced through his brain. "You're wondering what the autonomous unloading system will 'think' when it 'sees' the scaffolding outside or in its way? After all, it was designed to start unloading on its own—not with someone rebooting it from inside."

"Well, that was one thought that crossed my mind," answered Commander Carrigan.

"Not only that," Carl said, "but we'll be inside when we reboot it. Not that I think that's much of a big deal. I suppose it'll just start to boot up, which will take a few minutes, but it should 'see' the scaffolding and avoid it during the off-loading process or pause until the unloading zone is clear."

"I was wondering if we could simplify the scaffold to be more like a giant ladder once we have *Starship VII* off-loaded. After all, we only need to go up once and reboot the remaining starships. I mean, you did an absolutely incredible job with this wonderful scaffolding structure. But you understood during the construction of it that we were going to go up and down many times. Once we finish with *Starship VII*, all we need to do is go up once and get out of the way, probably relatively quickly."

"Oh, sure, I can do that," Carl said. "All I have to do is cut away some of the beefier sections of the scaffolding and reinforce the ladder portion. No problem. It'll make the transport of the ladder a lot easier, too, since it'll weight a lot less. I could put some 'legs' on it that will support the starship's ladder and keep it from bouncing so much while we ascend it. And that's all it'll need. One person goes up, boots up the computer, and we take it down." Carl was still thinking about the job at hand. "I bet I could reassemble the ladder idea in as little as a day."

"Do you have any other ideas?"

"I have a lot to do with all of this construction on the surface, yet we need service on the hydrogen power cells as backup to the solar system and the water system. I'm being pulled away from my normal maintenance routine, and I could use some help with *Staten's* day-to-day operations."

"Of course," the commander said. "What do you need?"

"Frankly, I need two of me." He shook his head, making his helmet bob from side to side. "I'm afraid if anyone else touches any of these sensitive systems on *Staten* and screws them up, we could all be in for a world of hurt. We either get water, or we get the starships up and running, or we run out of O_2 . . ."

Commander Carrigan suddenly realized how hard he was pushing Carl. None of this work on the Martian surface was in the protocol or plan or in Carl's job description on top of keeping up with the all maintenance regiment was really more than any human could handle.

"Carl, you've been doing all of this work with three or four of the same crew, right?"

"Yeah, pretty much" Carl said as he sat down beside the commander.

"Well, they know how to weld and use the cutting torch by now, don't they?"

"Yeah."

"Well, let them do what you taught them. I've been listening to you on the radio. You're great at teaching and instructing. Let them do the work on the surface so you can get on top of the demands of *Staten*. Tell them what to do, check in with them from time to time, and inspect their work. I'm sure they can do it."

"You're right," Carl said. "They can do the work while I get caught up on *Staten*. I want James and Wayne for sure. I know they're on different shifts, but I trust them."

"You got it, Carl. Which starship should we attempt this first reboot procedure on?"

Carl considered his options. "I think we're okay on O_2. We'll be okay on water once we get the water line in place and after I attend to the hydrogen power cells. This is kinda like Christmas," he said with a smile. "I think we need another Mars car. I am so tired of swapping personnel with limited transportation. So, whatever starship has the third Mars car loaded inside, that's the one I suggest."

Commander Carrigan had to laugh. "I think that's a great suggestion, and one that I support. That'll be our next goal. Now, let's see how far they've gotten to getting us water. I don't know about you, but I'm getting pretty thirsty."

As Commander Carrigan and Carl returned to *Starship VII*, they saw the crew jumping up and down. They had both forgotten to switch back to Alpha frequency. They hailed Pedro and instructed him to recognize that they were changing frequencies.

Pedro confirmed, "Commander Carrigan and Carl back on Alpha."

Commander Carrigan and Carl switched over to Alpha frequency to hear, "Yes!"

"Well, bring it down!"

"Give us a second to get it out."

"Do you really have it?"

"Yeah, I think so."

"Commander Carrigan on Alpha. Sorry, guys. I was having a conversation. What do you have?"

"Commander Carrigan, this is Haratu. We're bringing it down now. We have the water line."

As Commander Carrigan was about to instruct them to attach it to the crane, Haratu started climbing down the scaffolding ladder with some of the water line over his shoulder.

"No, put it on the hook!" shouted the commander. "Don't try to manually bring it down!"

Haratu swung his foot over the ladder, and the hose shifted its weight on his shoulder. The momentum pulled his arm and shoulder away from his clutch point on the ladder, and the heavy hose won the battle of momentum as it swung across his back, leaving him with one hand on the ladder. His left foot slipped off, and he flipped end over end with the water hose entangling him on the trip to the surface. He landed upside down with his untangled arm outstretched, his head somewhere between his legs, and his feet slamming down on top of him. His body bounced off Mars's surface like a rag doll thrown down by a disappointed child.

Everyone froze. It was obvious that Haratu was dead. Right in the middle of success and happiness was death; failure due to lack of consideration of the possible. The exhilaration of the moment was suddenly destroyed by catastrophe.

Commander Carrigan immediately hailed Pedro to activate flight surgeon Spencer and to have her ready for a trauma alert.

Why didn't he tie in with his safety belt? Commander Carrigan thought, damnit!

Per protocol, the whole team scrounged for a backboard, staged at all extended work scenes. They applied C-spine stabilization using the duct tape, followed by rapidly loading Haratu onto the nearest Mars car. He was not conscious, and he wasn't breathing. His body no longer took on a natural shape.

On Mars there is no way to monitor a trauma patient in a Mars suit. You cannot stop blood loss other than wrapping the location with duct tape. CPR was performed for the duration of the trip, but none of that would have mattered for Haratu Suzuki. His internal injuries were beyond paramedic or emergency surgical intervention.

What was left of Haratu was taken aboard *Staten*. Tammy Spencer met them upon equalization to *Staten's* climate, and she checked for life signs. It had been twelve minutes since the fall. She confirmed he was dead. No heartbeat, no respirations, and blown pupils confirmed brain death.

As per protocol, his body was removed from *Starship X* and taken to a remote hillside location, where they dug a grave in the Martian soil. His body was placed next to the grave and left in state, still in his Mars suit. Still with a job to do, Commander Carrigan instructed those still on the surface to connect the last of the water line to *Staten* and to turn on the pressurized water tank from *Starship II*. He then asked two members of the crew to walk the line from *Starship II* to *Staten* and confirm integrity of the line.

Despite the fall, the water line was intact, and soon they heard water filling the tank. What should have been a momentous, joyful occasion was met with a depressing acceptance. Commander Carrigan announced to no one's surprise a full-staff meeting immediately.

"We are here tonight to mourn and give thanks. Haratu Suzuki died today in service to our cause. Every one of us knows the risk we all took every day just to get here, not to mention being on the surface of this planet without air, without any significant atmospheric pressure, without water. Haratu was a wonderful person to work with. He was always full of energy and enthusiasm, which may have been his untimely demise. He was so excited to have located the water line that he misjudged its weight on his shoulder, and this excitement led to his last act. He was a true astronaut's astronaut, a trooper amongst troopers. He was my friend, as are all of you." With a whimper, the commander showed his humanness, and added, "I will miss him. I will miss his smile and his willingness to participate. It didn't matter the type of task or job, he was always there willing to help. I want to dedicate this water line to Haratu Suzuki. He helped to save our butts. Until we leave, this will be the HS water line, dedicated to an astronaut's astronaut. If Haratu were here, he would say, 'Never give up, never stop trying. Imagination leads to discovery, imagination leads to us, being here on Mars. You people are here because of imagination. You people are here because you are just like me. You people are of the same substance as he was and as he is through you! You all share the same energy, the same focus. You all know the pressures and tensions of getting through the relentless interview process, yet we all made it. Let us spend the rest of this evening talking about Haratu, his focus, his dream, and the times we spent together. We all dream the same dream— exploration and discovery—which is driven by imagination and courage. Nothing will be changed by Haratu's death. Nothing! It was a terrible accident. Haratu died, and he would want us to prove that we can survive on Mars. Not just survive, but thrive! And we will! And we are! So, I say thank you, Haratu, for your service to our cause. I say thank you, Haratu, for your attitude. I say thank you, Haratu, for your companionship. I say thank you, Haratu, for your love of our future. I say thank you, Haratu, for your imagination. I say thank you, Haratu, for your courage to help make it a reality. Today, tonight, we start to live on Mars. Because of you, Haratu. Because of your energy, we are here today. Please enjoy this water Haratu made possible. I love you, goodnight."

37. Johnson Space Center - Day 15

Brad Brown spent the rest of the evening going back over his plan if his messages were to get through to *Starship X*. Just knowing they had a way to turn on the other starships made him feel a little better, even though he knew they only had two 22-foot ladders, which would not get them close to the hatches on any of the starships.

Brad tried to imagine the conditions on board *Starship X*. If they were able to start up the MOXIE for oxygen and the Kilopower unit, they might still be alive. If they didn't, they were probably all dead by now. Just how would they get to the hatches? He kept going over the inventory on *Starship X* to see if they had anything they could use. Maybe the rope? They had a lot of rope. Could they lasso the top of one of the starships? No. No matter what he thought of, he couldn't imagine any way to get to the hatches. Except what his crew came up with – knocking the starships over, he couldn't see any other way. What a mess that would be if they had to do that. He was getting depressed.

His team continued in their attempts to initiate the starships by boosting the signals and sending them from different antennas. They had no way of knowing if any of their transmissions were getting through, but they certainly weren't getting anything back.

Brad listened to the 11:00 news while he had a bite to eat. The president announced the opening of the New York Stock Exchange as a test the next morning with emergency stops ready to be imposed. Anything that started to move excessively would be frozen. The local TV reporters were commenting on the old-fashioned cable system installation from the 1980s, with modern conductivity and technology upgrades.

Food had been slowly returning to grocery store shelves with forced "government guarantees" of payment in place. People were using checks, cash and the old credit card impression machines were back again. The banks were forced to return to paper documentation. It was working—just slowly. The civil unrest of the previous two weeks was beginning to simmer down. Everyone wanted to know who blew up the satellites.

The news was following a presidential and international investigation. Many countries had been thrown in the exact same situation, everyone was looking for the guilty party. All of the security agencies around the world were pooling their energies as well as running simultaneous investigations on their own. There were terrorist-plot theories, Russian theories, Cuban theories, and even conspiracy theories that the United States did it to gain a new economic high ground.

The press seemed to be as perplexed as everyone else, coming up with inexplicable theories on theories. All of it was conjecture with no factual evidence.

Unbelievable what BS the press can come up with, Brad thought.

People were beginning to return to work. It was as if the entire country had a bad hangover and was still suffering from a headache, but still putting one foot in front of the other. Brad felt the same way. He just couldn't see a happy ending in sight.

It was almost two o'clock in the morning, so he decided to go home and get some sleep. He guessed that the general must have gone home before getting his messages, or he had nothing good to report.

Brad's phone rang, and he jumped up and checked the number. It was blank. It rang again. "Hello? Er . . . um, Brad Brown," he said, trying to get a handle on his emotions.

"Brad! It's General Latimore."

"Hello, General. You're keeping some late hours," Brad said, not knowing where the general was at the moment." "Hello, Brad. I just landed at Edwards. My personal assistant leaves my messages on a voicemail system I can access when I'm on one of our bases. Anyway, I bet you'd like some feedback on your message."

"Yes."

"It turns out your theory was correct. The crew from the ISS was in communication with *Starship X*, and they were in communication with them for some time. I'm not sure how long, but it sounded as though they'd been in daily contact for maybe a week. They did get the message you provided, and they relayed that they felt the information was important for their current conditions. *Starship X* crew reported to the ISS that they had run out of water, but the information was of significance and could improve the situation. So, maybe the information you supplied helped. I'm not sure, but it sounded like they had just recently run out of water. At least they didn't carry on about it in distress. The personnel they usually spoke with on the radio was on the surface of Mars, addressing the situation. My details are a little sketchy because I didn't have much time with the crew. They were tired and stressed. I'm sure you'll get more time with them at a later date. Suffice it to say, your message was sent and received, and they seemed to think it was important, significant and welcome news. Brad? Hello?"

"Yes, yes, sir. I'm just trying to put this together in my head. Thank God they were able to get the message. Thank God they are still alive! It's going to take me a bit to digest this. We didn't even know if they were still alive, General. Very good news, sir," Brad said with some relief. "Do you have any more information or—"

"I'm afraid that's all I have, Brad. I would like to congratulate you on a job well done. You made a difference today. I have to tell you that none of us thought that message would get through or that they were actually in contact with the crew on Mars. Good job."

"Thank you, sir. I do appreciate it. Hopefully, it will help."

"You're welcome, Brad. I'm sure you'll be able to get in touch with the crew once they return stateside and get a better debrief. But you can rest assured that they did get the information. Once again, congratulations on a job well done. We will talk later. Good night, get some rest. Oh, and Brad, the crew was pretty tired and groggy, but one of them repeated something a couple of times. He said, *'Aut viam inveniam aut faciam.'* I don't remember much of my Latin anymore, but I thought I would pass it along."

With that, the general was gone. Brad sat back, half-mesmerized by the conversation. They were out of water. That probably means they hadn't refurbished the hydrogen fuel cells. They can't survive on the hydrogen fuel cell's byproduct of water alone, but it would help.

Brad stared blankly at a wall across the room. I wonder what they're up to. They must be trying to get the remaining water line for the run from *Starship II*. How much water line was on board *Starship X*?

He quickly pulled up the packing manifest for each starship. Hmm, yeah, 3000 feet. Why didn't we put enough water line on *Starship X* for the run from *Starship II* to the landing zone for *Starship X*? Because the starships were supposed to be completely off-loaded on the first day, as soon as they landed. That's why. No one expected the starships not to autonomously off-load. What else is going to happen that we didn't expect? What did the general say? *Aut viam inveniam aut facium*? Brad looked it up on his computer. Nothing. In his office upstairs, he had a handbook of Latin sayings. He ran to the room and pulled out the book.

Aut viam inveniam aut facium.

I will either find a way or make one.

38. Mars - Sol 15

Before the night shift was off and before the day shift started, as everyone expected, Commander Carrigan called a full-crew meeting for the review of the accident leading to Haratu Suzuki's death.

The results of the description of the accident were simple. Haratu Suzuki made a poor decision because he was excited to have found the water hose. His emotions overwhelmed his decision-making process, and it led to an unintended and unexpected outcome. Commander Carrigan reiterated that the dangers of the job were magnified by the fact that they were way outside the original parameters of their designed constraints. No one expected to have to climb a 150-foot scaffold on a daily basis. Yet they still had to continue to climb the scaffold until all of the starships had off-loaded their supplies.

The situation was more stable now that they had water and their O_2 was holding. Commander Carrigan reiterated the first accident injuring Pedro was caused by rushing to get the job done. Overzealousness and understandable excitement caused both accidents.

"Let's be careful out there, people. Let's not let our emotions control our actions. Does anyone have anything they want to add or contribute?"

No one said a word.

"I'm sure you've heard by now that the ISS gave us the instructions on how to boot up the autonomous systems on the other starships. We will still have to climb the precarious scaffolding to engage the system since it cannot be done from the ground. Also, that was the last message from the ISS. They evacuated the Space Station just as it was becoming uninhabitable. Apparently, our instructions were sent to them on a heavily reinforced Crew escape capsule. MMI or CAPCOM was hoping the ISS was in communication with us. Pretty good thinking back at CAPCOM. Those of you that were on A shift can get some sleep. B shift will be working on off-loading *Starship VII* for four hours. Does anyone need Carl working with them for instructional purposes? He's been out there relentlessly for days on end, working way too much and way too hard."

Again, no one responded.

"Please be careful, people. It is twenty hundred Zulu. B shift, get something to eat, and suit up at twenty-two hundred. C shift will replace you at oh two hundred until oh six hundred. Take as long as you need to do your job safely. If anyone feels tired or not up to the job, let me know. I don't care if it takes three days to finish unloading *Starship VII*. I want all of you safe. For *Staten's* use only, I will be adopting a Martian Time Zone for *Statens* use only, since Zulu will be moving a whole day ahead of us soon. All comms for Mars surface use only. Any extraplanetary times remain on Zulu if we had any communications, as well as all documented reports remain in Zulu. Just refer to the computer for Zulu time. Otherwise, *Staten's* time zone will be the official time on Mars from now on. If anyone needs to talk to me or needs personal time, I will be in my office. Let's be safe out there." Commander Carrigan grabbed a cup of coffee. "Carl, can I see you in my office, please?"

Commander Carrigan intended to check on the O_2 levels, but he wanted to talk to Carl, and he knew Carl knew what he was interested in finding out. "Once again, I'd like to get an update on *Staten's* vital signs. Did you get a chance to check on the water tanks and the O_2 levels this morning?"

"Yes, sir. Those are the first two items I do every morning."

"I know. I know, you're a good man Carl" the commander responded with a smile.

Carl looked up and smiled back "The water tank is full, and our O_2 has been stressed out due to all of the activities and EVAs over the past few weeks. We're holding steady at a ten-day level of oxygen reserve, but the MOXIE is working, and I expect our reserves to rebuild slowly."

"That's what I figured as well. How are your energy levels? You've been so busy lately."

"I'm okay. I'm just glad we have water now and that the O_2 levels are no longer at emergency levels. I'm impressed the batteries worked as well as they did considering how little the hydrogen fuel cells were producing before we set up the Kilopower unit. Which is once again powering the MOXIE, as well as *Staten*."

"I agree. What do you have on your schedule for today now that you don't have to help with the off-loading?"

"Well," Carl said, "I have to get to those hydrogen fuel cells that need servicing. They're a good backup if the Kilopower unit goes down. Then I need to get caught up on my normal maintenance routine, which I've been getting way behind on. We're going to have to deal with our garbage levels. I suggest we find a location to put it on the surface or bury it. Also, I have an idea for the dirty towels that are building up. I was thinking that since the sun's radiation kills pretty much everything , if we put them out in the sun—like on an old fashioned clothesline—it would kill the bacteria and remove a lot of the smell."

"Yeah, we are going to have to give some consideration to recycling them. I'll ask Barbara if she has any ideas, but I sure like your clothesline idea. It makes sense and at least it gets them out of *Staten*. I sure wish we had a washing machine."

"Hmm, yeah," Carl said. "I would love a normal shower."

"Wouldn't we all. Is there anything else that needs to be brought to my attention? We've been so busy, I don't want any rude surprises related to crew safety, security, or quality of life."

"No, I think we're secure at the moment, and I don't see any cause for concern on any life safety systems or internal environmental conditions. I do need to go over my extensive service logs to review every aspect. I don't see anything surprising us at the moment."

"That's really good news, Carl. Thank you for the update."

After Carl left, Commander Carrigan had some time to depressurize and think about all of the situations that had occurred over the past weeks—and what their future may hold. The first humans to walk on Mars. The low oxygen situation that threatened the entire crew. Running completely out of water. James Galway and the pallet of supplies that fell out of *Starship IV* and badly injured Pedro. Haratu Suzuki falling to his death. Discovering water and life on Mars in the cave on the same day, only four miles from *Staten*. Loss of communication with CAPCOM. Eleven months left to finalize their mission on Mars. What else could the cave be concealing for future discovery? How much farther down does the cave or lava tube go? Could they use it for protection in the case of a significant solar flare? Will they ever know before a large solar flare is released from the sun? There must be more caves like that one around. Do they have life living in them as well? How common is there life in these caves? Can Barbara and *Staten's* crew get food to grow in the soil from the cave? Is the water in the cave fit for human consumption? Is it a replenishable resource? Will they be able to return to Mother Earth after the eleven months? How will they know when it is safe to return? It appears the Kessler Effect has taken place. How will I tell the crew what that means? Will the Kessler Effect ever get cleaned up enough for them to return? How will the crew react if we can't return home? Will we be marooned here on Mars for years? Could we end up being the first Mars colonists? What will we do for food? Can we grow food here? How long can we last? How long will the MOXIE last or the Sabatier equipment?

These were just some of the issues Commander Carrigan needed to consider and prioritize for the future of *Staten* and the first humans on Mars. Mars Mission I had more twists and turns than anyone could have expected and they had been on the surface of Mars for only 15 days.

The commander was shaken from his thoughts by a sudden yelling and pounding on his door.

"Commander! Commander!" Steven rushed in without an invitation excited and out of breath. "They're doing it!

"Doing what?"

"The starships just started up on their own. They're off-loading their supplies on their own! They're working!"

39. Bob and Brian

After Bob saw the missile take off, and even before they fully realized what had happened, they were being hailed by a very aggressive, fully dressed-out Coast Guard crew. The machine gun on the bow of the response boat was fully loaded and manned by a guy looking like he was going to war. And it was being aimed at Bob and Brian. The Coast Guard landing party of four enlisted men and one officer covered head to foot in camo and bullet proof vests pulled up along-side their boat, shouting orders at them from a rigid hull inflatable boat. Bob had been attending to the cut on his incapacitated partner's head and didn't realize the Coast Guard was shouting orders at him. His ears were still ringing from the rocket launch.

Coast Guard Station Fort Lauderdale crew members weren't happy with them. Bob hadn't returned the calls on the radio or acknowledged the loud hailer mounted on the front of the response boat. They could see he was on the deck of his boat, but they didn't know what he was doing. This made the small fishing boat and its two occupants extremely suspicious.

The two men were below the freeboard and the railings on the deck of their boat, effectively but not intentionally, hidden out of sight, and not responding to multiple hails or to repeated attempts at communication. The Coasties simply didn't know what to expect, so they were prepared for anything. It was a good thing Bob didn't stand up unexpectedly with his fishing pole in his hand or anything else for that matter. The Coast Guard crew had their hands on their sidearms as they rapidly boarded the twenty-seven-foot catamaran and surrounded the two men.

As he was being hand cuffed with the zip ties and taken into custody, Bob yelled, "What do you guys want?"

The Coast Guard officers began to realize that Bob was deaf, and Brian was borderline conscious, if that. They were able to get Bob to quiet down as they started asking an endless line of questions. None of which Bob could hear. Finally, using a whiteboard they had back on the larger, forty-five-foot vessel, they were able to get some answers to their questions. Background checks were performed, and their stories matched what was known about them. The fishing vessel's registration was current and it was inspected from bow to stern and put in tow behind the forty five foot Coast Guard Response Boat Medium. Because they had both been firefighters, their ID's matched their fingerprints which had been entered into the federal databases when they received their State of Florida Fire Fighting Certifications and they seemed to be exactly who they said they were and doing exactly what they said they were doing - fishing. They were released from the zip tie cuffs.

Brian wasn't much help to the interrogating officers. He appeared to have a concussion, and the Coast Guard officer in charge called for Fire Rescue to meet them at the dock. A much more empathetic Coast Guard EMT applied a new bandage to Brian's head after inspecting the injury which was swelling up like a hard-boiled egg on the side of his head.

Bob and Brian were delivered to Fifteenth Street Marina, where a Federal Marshal accompanied them to Broward Health Medical Center in the back of a Fort Lauderdale Rescue unit. They would stay in custody for at least another two hours. After repeating their story to some forty or fifty people, seemingly from a dozen different federal agencies. They were finally cleared as witnesses and were reminded numerous times that the federal officers may be back if they felt like it.

Their boat was tied up outside the Fort Lauderdale police station and marina, where it would be inspected again by more federal employees. They even paid the local marina to pick the boat up in the air, so it could be inspected again even after the divers had inspected the entire bottom side at the dock. The doctors wanted to keep Brian overnight, but he insisted that he was okay and would check in with his personal physician in the next few days. When the Dr.'s heard both of their stories and watched the stream of Federal employees coming and going they were rather happy to let them go home.

Bob had an old television antenna in the attic, which was able to get some local stations. Only four local stations were airing, providing national news coverage.

Bob shook his head and took a swig from his beer. "Man, Brian, that seems like such a long time ago already."

It had been fifteen days since the missile launched right in front of them, when they were the sole witnesses of an act of war against the United States and against the rest of the World as far as that goes. Everyone had lost satellite communication, cell phone and commerce, services and benefits. As far as they knew, no one else had seen it. Each time they had power again and the news came on, Bob and Brian were glued to the TV. They wanted to get as much information as they could about the stranded twenty one astronauts on Mars. They felt unusually attached to them, as if they were somehow now a part of the mission.

"This is so bizarre. We actually watched *Starship X* launch from Pad 39a six months ago, and now NASA has lost all communications with the crew on Mars. We've lost all of our satellites, the internet is gone, and the whole world has changed as a result. And we watched it, Bob."

"I sure hope they're all right up there all on their own," Brian said. "We saw some serious shit in our day. People in serious trauma, run over by trains, jumped off the top of buildings or bridges, cars overturned, drownings, and some rip-roaring fires, but we normally go home after twenty-four hours with no issues. Those astronauts are gone for two years, and now the news says they can't get to their food supplies, they could be out of water or low on air. They must be freakin' out." Could you imagine being thirty five million miles away from home and be out of air?"

They listened to every detail, like they were a part of history and they were.

Brian's head was healing up quickly, and his memory seemed to be returning to normal. He recalled seeing something floating in the ocean off of the port bow, but everything else was still fuzzy.

At first, when they told their buddies at Station Eight what had happened, no one believed them, no one even said a word. They all thought they had both been into the beer a little too much that day.

Even Battalion Chief Delany stared at them in disbelief. "Why weren't you guys on the news?" He said incredulously.

"I guess because there was so much news that day," Bob said. "We've both wondered that in the days since it occurred, but the press must've been overwhelmed, and the Feds were not interested in attracting attention. There was a reporter and a camera guy at the dock, but they probably couldn't get much from the Coast Guard or the Federal Marshals and they ended up dropping it, we guess. We went straight from the Coast Guard boat to the Rescue unit which was parked right next to the dock. And you guys know none of them are going to talk. And we've had all the attention we need for a while."

Chief Delany kicked back in one of the dayroom chairs. "Well, at least you guys made it back safe and sound. It sounds, well out of this world. Fortunately, we're off the generator, the landline phones are working our radios kept working the whole time, thank God, and we're back to a more normal and regular shift schedule . . . for now. We went two weeks twenty four hours a day non-stop with everyone called in for all shifts. No one ever went home. It was crazy! Worse than hurricane Andrew! Thank God you guys are safe and were no closer to the launch than you were. You could've been burned beyond recognition." He stood and gave them both a big hug. "We are, however, still in a state of war with whatever asshole did this, and there will be ramifications. There will be fallout. Trust me. I just hope it doesn't come back to haunt us. We have already had enough of this craziness!"

40. The Oval Office - Day 16

"Mr. President, we believe we know who was behind the attack," reported Secretary of Defense Robert Thornton.

President Gerald Gary Griffin preferred his friends to call him Gary, was referred to by some as "The Triple G," and was not a man to be taken lightly, as was witnessed by his aggressive presidential campaign and subsequent time in office. He had been on the phone with prime ministers and presidents from Russia, the UK, Canada, Japan, France, Germany, China, India and more. They all had ideas, but no one had facts. Now his own secretary of defense told him they believed they knew who it was.

President Griffin was a facts man, and a statement starting with "we believe" belonged in church, not in *his* office.

"What do you mean you *believe*?"

"Sir, we have a high level of confidence in our information," said Secretary Thornton. "The odds of finding the offending party in any crime are reduced with time. Due to the loss of our satellites, we cannot say with one hundred percent certainty. We have not shared this information with any other government, nor have they shared it with us. But a simple deduction of motivation consistently returns us to one country, which has only recently developed the technical skill to pull off this type of action, and one of the few countries with anything to gain from the consequences."

President Griffin was listening. This was the first time anyone had come up with a theory. He hated theories, but it couldn't hurt to listen. "Go on."

"This is obviously a rare situation. Every developed country, and many undeveloped countries, is a victim. If you ask yourself what country could possibly stand to benefit from this, you look at countries that don't like satellite flyovers. Your first thought is Russia, but when you realize that all of their satellites were knocked out of commission as well, then you have to ask yourself: Would Russia shoot themselves in the foot to gain virtually nothing? One could easily see that Russia is worse off after the attack, just as we are, so it wouldn't make any sense for them to initiate it. Their economy was hit as bad as anyone's. If you look at it purely from an economic standpoint, you might say that only third-world countries weren't affected, but that isn't true either. We have years of history of trading with countries such as Vietnam, Cambodia, Bangladesh and many more. After the incident, all GPS satellites were lost, as were all banking

337

transactions. And all trading with these countries has stopped—not slowed, stopped. The money flow has stopped. If their ships loaded full of goods for the United States could sail, they would have to navigate by the stars like in the 1800's. And they couldn't off-load their supplies until they are paid in cash. Who is going to meet them at the dock to pay millions in cash, so they can off-load their goods? They have literally been isolated from the money supply. We keep talking about how bad it is here, but Mr. President, some of these countries are looking at serious famine and possibly revolution. Backing up just a bit, you might say yeah, but none of these countries possess the kind of technology to initiate this attack and not be significantly injured from the effects. And that is the key. When you go down the list, there's only one country that doesn't rely on their own satellites or at least very little. There is only one country that was predominantly untouched by the implications of this attack, yet has the technology to carry it out, therefore, leveling the playing field. We know who it is. And if we know, it is only a matter of time before the other big time players know. If they don't already. So you need to decide how you want to play it."

41. Ryongsong Residence –

North Pyongyang, North Korea

Having just returned from another children's dance-and-drum presentation, Kim Jong-un smiled as he relived the presentations, reminding himself to authorize an extra food allocation for the children's families. They worked so hard to please their Great Leader.

On his desk were the notes from his personal assistant. He had received an official message from the Norwegian Embassy, which wasn't surprising, but it was strange that the diplomatic courier hadn't stayed to personally deliver it. The envelope itself had gone through the normal security checks, so it was a bit odd it had three different seals on it, demonstrating an unusual amount of security for normal communication with Embassy personnel.

Still taking every precaution, Kim Jong-un had his personal assistant open the envelope, which contained another sealed envelope inside. She opened the second one as Kim Jong-un looked on. Inside that envelope was a third sealed envelope.

Kim grabbed the envelope from his assistant and tore it open impatiently. In the middle of a single piece of folded paper was a typed message in French: NOUS SAVONS.

We know.

His personal assistant was watching for an emotional reaction to this rather strange note. "The Russians showed up unexpectedly and are in reception, she said quietly, still watching him for a reaction."

The paper slipped silently from his hands and drifted to the floor.

Christopher Lee Jones grew up in Southern Michigan, attended Jackson High School and Michigan State University. He moved to South Florida just before hurricane Andrew, became a scuba diver and obtained his USCG Captains license working in the Merchant Mariner industry and he helped with many deliveries of boats around the Caribbean. He worked at Miami-Dade Fire Rescue for many years, retiring in 2019. A self-admitted space nut in hibernation, he witnessed one of the first Falcon 9 launches when the booster returned and landed safely at Kennedy Space Center. He also attended the first Falcon Heavy launch when Spacex sent Elon's red Roadster into space and both boosters landed safely back at Kennedy Space Center. He then knew he needed to pursue his love of Space. He has since traveled to many NASA destinations including the U.S. Space and Rocket Center in Huntsville, Alabama (a must see), has toured the Marshall Space Flight Center and the Redstone Arsenal, where the staff patiently answered all of his numerous questions. Today he resides in South Florida and continues his search for anything Space related and how to return the 21 astronauts on Mars Mission I safely back to Mother Earth.

Made in the USA
Columbia, SC
18 April 2023

15536860R00187